AF697346

GENETIC ENGINEERING

Edited by **Idah Sithole-Niang**

Genetic Engineering
http://dx.doi.org/10.5772/46188
Edited by Idah Sithole-Niang

Contributors

Richard Mundembe, Youjia Hu, Mariana Ianello Giassetti, Fernanda Sevciuc Maria, Mayra Elena Ortiz D'Ávila Assumpção, José Antônio Visintin, Geert Angenon, Miguel Alcíbar

Published by InTech

Janeza Trdine 9, 51000 Rijeka, Croatia

Edition 2014

Copyright © 2013 InTech

All chapters are distributed under the Creative Commons Attribution 3.0 license, which allows users to download, copy and build upon published articles even for commercial purposes, as long as the author and publisher are properly credited, which ensures maximum dissemination and a wider impact of our publications. However, users who aim to disseminate and distribute copies of this book as a whole must not seek monetary compensation for such service (excluded InTech representatives and agreed collaborations). After this work has been published by InTech, authors have the right to republish it, in whole or part, in any publication of which they are the author, and to make other personal use of the work. Any republication, referencing or personal use of the work must explicitly identify the original source.

Notice

Statements and opinions expressed in the chapters are these of the individual contributors and not necessarily those of the editors or publisher. No responsibility is accepted for the accuracy of information contained in the published chapters. The publisher assumes no responsibility for any damage or injury to persons or property arising out of the use of any materials, instructions, methods or ideas contained in the book.

Publishing Process Manager Ana Pantar
Technical Editor InTech DTP team
Cover InTech Design team

Additional hard copies can be obtained from orders@intechopen.com

Genetic Engineering, Edited by Idah Sithole-Niang
p. cm.
ISBN 978-953-51-1099-6

INTECH

open science | open minds

Contents

Preface

Genetic engineering as a field emerged with the dawn of molecular biology when James Watson and Francis Crick deciphered the three dimensional structure of DNA. Several decades have passed since then, and today whole genome sequences of several different organisms across kindgoms have been deciphered giving rise to a wealth of information that is only a computer terminal away.

The book, while having only five chapters ,covers a wide range of topics in genetic engineering of microorganisms, plants and animals. Specifically, it covers both the natural and social sciences. In the natural sciences topics ranging from the genetic engineering of microorganisms to produce antibiotics, the gene targeting and transformation in plants, the generation of marker-free plants in response to biosafety concerns, as well as the generation of transgenic animals and those derived through cloning are covered. In the social sciences, the issue of ethics in biotechnology and the role of the media in reporting around the cloned sheep, Dolly are discussed.

Interestingly the application of genetic engineering to plants and animals generates controversy compared to the same application in microorganisms. Perhaps some of that controversy is due to the actual use of the end product, in the case of plants, food, and in the case of animals, even more so, as the reality of cloning animals conjures up fears that humans are about to be cloned even though cloning is not a transgenic process. This then brings on the whole issue of ethics in science, given the technological leaps and bounds of genetic engineering within the last decade compared to many before that. The book also highlights the role of the media in perpetrating the fear associated with the unknown. Clearly the role of the media should be one of responsible reporting with the goal of educating the public and not that of engendering fear.

The first chapter of the book discusses the genetic engineering of Acremonium chrysogenum, a producer of the antibiotic cephalosporin C, as well as 7-amino cephalosporanic acid a key intermediate for the biosythesis of many first-line antibiotics in this class. This paper is therefore pivotal to the manufacture of newer and improved antibiotics and certainly broadens the scope of this publication.

The second chapter deals with strategies for gene targeting using two specific examples of common selectable marker genes (SMGs), antibiotic resistance genes and herbicide resistance genes. They state how their residual effects affect the final product, and susbsequent use. They discuss the likelihood of the SMGs being able to spread to nearby crops, weeds or wild relatives. They then go on to review the various approaches currently used to generate marker gene-free plants, while highlighting the limitation of each method. Of particular importance was the the need to have these protocols widely applied to include sterile plants, vegetatively propagated species, as well as their application to plants with a long life cycle such as trees. They also described the improvement of these approaches as applied to both nuclear and chloroplast transformations systems. They conclude by showcasing the latest improvements as applied to agronomically important traits and most importantly the development of recipient plants that can be used to accelerate the breeding cycle and address the biosafety regulatory issues. At least from the genetically modified organism debate point of view one would hope that concerns with SMGs will be put to rest and aid adoption of the technology while recognizing that this issue is complex.

The third chapter is slightly different from the one above in that it discusses a slightly different approcah to gene targeting and genetic transformation. This approach has largely been led by controversies in GM crops. To-date this approach does not render the plant marker free per se but does begin to address some of those issues. This approach also provides a mechanism for producing engineered plants that might be non-transgenic depending on the source of the gene. It is routinely applied in Drosophila, yeast and mice and indeed in industry where made to order organisms are routine. The fundamental approach for animals is via homologous recombination (HR) whereas in plants HR does not work, and what seems to work is non homologous end joining (NHEJ) or illegitimate homologous recombination. The author gives an overview of other recombination systems based on serine and tyrosine recombinases, then ventures into the latest application of zinc finger nucleases that are capable of editing the genome without leaving a trace of the transgenic event. They argue that when the system is up and running this would be one approach to generating marker-free, vector backbone free precisely engineered crops that would be useful to the farmer, consumer and the environment.

In chapter four the authors delve into the entire spectrum of recombinant DNA technology as applied to animal systems. Indeed this is a comprehensive coverage of the subject. For example with DNA sequencing the topic is well covered from the basic procedure through direct PCR, capillary, microarrays as well as sequencing by MALDI-TOF culminating in various approaches to gene synthesis. Similarly, with cloning vectors, the entire spectrum of vectors is presented starting from plasmid vectors, for cloning small fragments, animal virus vectors, transposons into the large capacity cloning vectors such as Bacterial artificial chromosomes. The fundamental difference between an animal clone and a transgenic animal is adequately discussed. The technical challenges related to both approaches are well elucidated. The chapter culminates in the ethical consideration of genetic engineering. This is perhaps where the topic got a limited coverage as it only makes quick reference to issues of public perception and the ethics of the technology. Needless to say, I think giving detailed coverage at this level goes beyond the scope of the chapter.

In the fifth chapter the author argues very convincingly that during the coverage of Dolly the sheep, a lot of misconceptions were made, that the media set the scene and went to report the findings in a biased way. They further argue that the focus was on the culmination of Dolly and the social consequences of the experiments and not on the technical details. There were arguments as well around the benefits that accrue to both scientist and the techno-scientific companies. These benefits range from gaining professional prestige, legitimacy as well as financial gain. The mediatization of Dolly engendered fear in the public, resulting in the kinds of divergent views we see surrounding this debate today. Furthermore, the skewed intepretation by journalists fueled the debate on ethics with little attention paid to the actual science, such as the validity of the data or methodologies used. The consequence of all this was flawed decision making at both the administrative and legislative levels, with far reaching consequences on how the public view cloning.

Finally, I would like to express my sincerest gratitude to the authors of the chapters for the tremendous contributions. I am equally indebted to the Publishing Process Manager, Ms. Daria Nahtigal initially, Senior Commisioning Editor, and Ms. Ana Pantar, at InTech.

Idah Sithole-Niang
Professor, Department of Biochemistry
University of Zimbabwe

Section 1

Genetic Engineering in Microbes

Chapter 1

Genetic Engineering of *Acremonium chrysogenum*, the Cephalosporin C Producer

Youjia Hu

Additional information is available at the end of the chapter

http://dx.doi.org/10.5772/55471

1. Introduction

Acremonium chrysogenum, belongs to Filamentous fungi, is an important industrial microorganism. One of its metabolites, cephalosporin C (CPC), during fermentation is the major resource for production of 7-amino cephalosporanic acid (7-ACA), an important intermediate for the manufacture of many first-line anti-infectious cephalosporin-antibiotics, in industry.

Cephalosporins belong to the family of beta-lactam antibiotics. Comparing the first-discovered penicillin, cephalosporins have obvious advantages since they are more stable to penicillinase and are more effective to many penicillin-resistant strains. The incidence of adverse effects for cephalosporins is also lower than that for penicillins and other anti-infectious agents. Thus, cephalosporins are among the most-widely used anti-infectious drugs clinically. In China, the research on cephalosporins started from the 1960s, and cefoxitin was first developed in 1970. In the past 30 years, cephalosporin-antibiotics are one the most developed medicines on the domestic market. They accounts for more than 40% of the anti-infectious drug market share.

As the major resource for manufacturing 7-ACA, the production and cost of CPC is of the utmost importance in the cephalosporin-antibiotics market. The Ministry of Science and Technology of China has listed the fermentation of CPC as the major scientific and technical project in the past 30 years due to the continuous demand of strain improvement for the CPC-producing *Acremonium chrysogenum*.

Because of the limitation of traditional techniques on strain improvement for *A. chrysogenum*, along with the ubiquitous applications of molecular biology, genetic engineering has become a powerful tool to manipulate the antibiotic producing strain and to obtain a high-yielding mutant strain. This paper will summarize the most recent developments on genetic manipulation of *A. chrysogenum*.

INTECH
open science | open minds

© 2013 Hu; licensee InTech. This is a paper distributed under the terms of the Creative Commons Attribution License (http://creativecommons.org/licenses/by/3.0), which permits unrestricted use, distribution, and reproduction in any medium, provided the original work is properly cited.

2. Biosynthesis of CPC

The industrialization of CPC fermentation has been established tens of years ago with the breakthrough in key technologies including fermentation yield, fermentation regulation and preparation and purification. Nevertheless, there has been a lot of publications, recently on the improvement of CPC-producing strain by traditional methods, such as UV [1] or NTG [2]mutagenesis, and optimization of fermentation process [3], as well. However, most of the latest strain breeding techniques are at the molecular level, and the most important approach has been the research on the biosynthesis of the target metabolite.

The biosynthesis of CPC during the fermentation of *A. chrysogenum* has been well investigated. There are two gene clusters on the chromosome that are involved in the biosynthesis of CPC. The "early" cluster consists of *pcbAB-pcbC* and *cefD1-cefD2*. The *pcbAB-pcbC* encode two enzymes responsible for the first two steps in CPC biosynthesis [4]. While the *cefD1-cefD2* encode proteins that epimerize isopenicillin N (IPN) to penicillin N [5]. The "late" cluster consists of *cefEF* and *cefG* genes, which encode enzymes responsible for the last two steps [6].

The biosynthesis pathway of CPC is illustrated in figure 1. The ACV synthase, encoded by the *pcbAB* gene, condenses 3 precursors L-α-aminoadipic acid, L-cysteine, L-valine to the ACV tripeptide. The ACV is then cyclized into IPN by IPN synthase encoded by *pcbC* gene. The step from IPN to penicillin N is catalyzed by a two-component epimerization system encoded by *cefD1-cefD2*. The *cefEF* encodes a unique bi-functional enzyme, deacetyloxy-cephalosporin C (DAOC) synthase-hydroxylase which successively transforms penicillin N into DAOC and deacetyl-cephalosporin C (DAC). The last step in CPC biosynthesis is catalyzed by a DAC-acetyltransferase (DAC-AT) which is encoded by *cefG*. The crystal structure of DAC-AT has been published [7]. It has been shown that DAC-AT belongs to α/β hydrolase family according to the formation of DAC-enzyme complex [7]. Among these, *pcbAB*, *cefEF* and *cefG* were considered as the rate-limiting steps in CPC biosynthesis [8].

In recent years, some other regulatory proteins, which have been found to be important in CPC biosynthesis, as well as their coding genes have been discovered. For example, Ac*veA*, a homologue of *veA* from *Aspergillus*, regulates the transcription of all 6 major CPC biosynthesis genes including *pcbAB*, *pcbC*, *cefD1*, *cefD2*, *cefEF* and *cefG*. Disruption of Ac*veA* leads to a dramatic reduction of CPC yield.

A *cefP* gene located in the early cluster of CPC biosynthesis cluster has just been characterized. This gene encodes a transmembrane protein anchored in a peroxisome. It regulates the epimerization of IPN to penicillin N catalyzed by CefD1-CefD2 two-component enzyme complex in peroxisome. The *cefP* disruptant accumulated IPN and lost CPC production [10]. To compensate for the disruption of *cefP*, both *cefP* and *cefR* need to be introduced simultaneously. The CefR is the repressor of CefT, and stimulates the transcription of *cefEF*. A mutant *A. chrysogenum* without *cefR* showed delayed transcription of *cefEF* and accumulation of penicillin N resulted in reduction of CPC yield [11].

A *cefM* gene was also found downstream of *cefD1*. Disruption of *cefM* accumulates penicillin N with no CPC production at all [12]. It is suggested that CefM may be involved in the

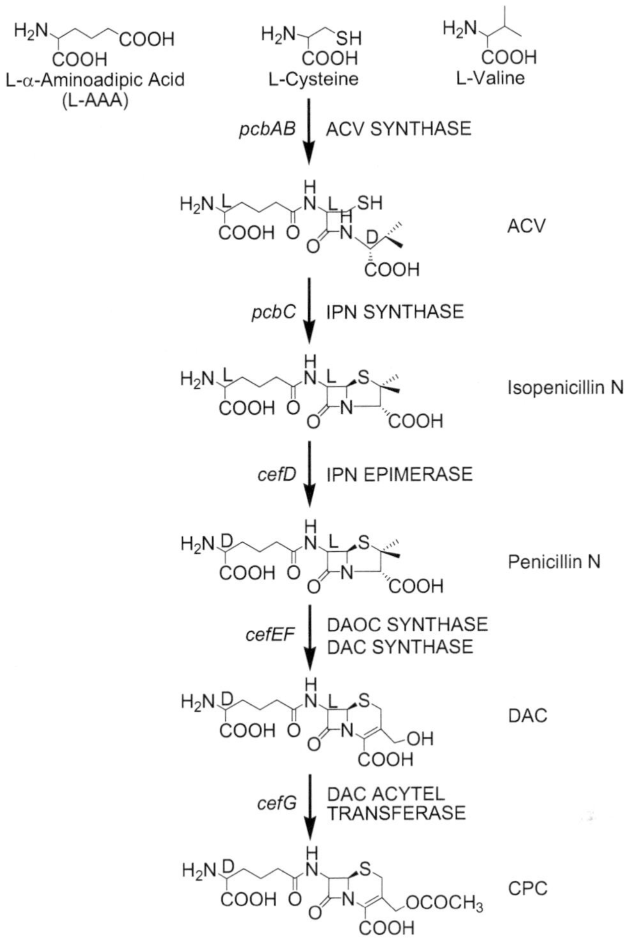

Figure 1. The biosynthesis pathway of CPC

translocation of penicillin N from the peroxisome to the cytoplasm. Without *cefM*, cells are unable to transport penicillin N which gets epimerized in peroxisome into cytoplasm, from where CPC is synthesized.

3. Techniques for molecular breeding

Acremonium chrysogenum belongs to the family of Filamentous fungi. The techniques for genetic breeding are somehow difficult to manipulate due to its complicated structure of the cell wall and the special life cycle. Our laboratory has started the molecular breeding of *A. chrysogenum* at a relatively early stage based on some published results from host, transformation, homologuous recombination and selectable marker of *A. chrysogenum* [13, 14].

To introduce exogenous DNA into *A. chrysogenum*, a traditional PEG-mediated protoplast transformation method is commonly used [15]. Since we are focusing on high-yield, or industrial strains, which usually have a stronger restriction-modification system than type strain, the traditional transformation method is not efficient enough for foreign gene introduction.

Agrobacterium tumefaciens mediated transformation has been widely used in plant genetic engineering, and in some of the Filamentous fungi including *Penicillium chrysogenum* and *Aspergillus nidulans* as well [17]. We have developed an adapted *A. tumefaciens* mediated transformation protocol for *A. chrysogenum*, which has a higher transformation efficiency than the PEG- mediated method [17], and more importantantly, this protocol can also be applied in *A. chrysogenum* high-yield strain. This is the first report of *A. tumefaciens* mediated *A. chrysogenum* transformation in the world.

Considering the significant improvement after introduction of *vgb*, VHb protein coding gene, we use error-prone PCR together with DNA shuffling to artificially evolve the *vgb* gene *in vitro*. After primary and secondary screening, a higher active mutant protein was obtained. *E. coli* bearing this mutant VHb produce 50% more biomass than its counterpart bearing the original VHb under limited oxygen environment [18].

A lot of basic research was done to facilitate the genomic DNA extraction [19] and endogenous promoter capture [20] from the chromosome of *A. chrysogenum*. A notable progress is the cloning of *pcbAB-pcbC* bi-directional promoter from the chromosome of *A. chrysogenum* [21]. This allows for the convenient manipulation of *A. chrysogenum* by introduction of multiple genes.

The last step in CPC biosynthesis, DAC transformed into CPC catalyzed by DAC acetyltransferase, was further investigated, as many reports have demonstrated that this is the rate-limiting step while DAC acetyltransferase coding gene, *cefG* has a low transcription rate *in vivo*. Our study showed that recombinant expressed DAC acetyltransferase can transform DAC into CPC *in vitro* in the presence of acetyl CoA [22]. The enzymological and kinetic study of the recombinant DAC acetyltransferase help us better understand the catalytic mechanism of the enzyme and make it possible to improve its enzymatic activity *in vivo* [23].

4. Molecular breeding of *Acremonium chrysogenum*

Among the three rate-limiting enzymes, PcbAB is relatively difficult to manipulate due to its larger coding gene. Thus, researchers focus on *cefEF* and *cefG* for molecular breeding of *A. chrysogenum*. Besides, extra copy numbers of *cefT* could increase the yield of CPC in the mutant *A. chrysogenum* [24]. And, overexpression of *cefP* and *cefR* in *A. chrysogenum* can decrease the accumulation of penicillin N and promote the yield of CPC by about 50% [11].

The fermentation process of *A. chrysogenum* is an extreme oxygen-consumption procedure. All the rate-limiting enzymes are oxygen-requiring enzymes. The *Vitreoscilla* Hemoglubin (VHb) is very attractive since it is capable of oxygen transmission in oxygen-limiting environments. A recombinant strain bearing VHb can significantly improve the usage of oxygen during the fermentation process and increase the product yield, which has been proven in *Aspergillus* [25]. Introduction of *vgb*, the coding gene for VHb, into *A. chrysogenum* can also maintain a higher specific growth rate and specific production rate resulting in a 4-5 fold higher yield of the mutant strain [26]. Actually, there are many industrial *A. chrysogenum* strains that express a recombinant *vgb*.

The earliest report on genetic modification for *A. chrysogenum* was published in 1989, when researchers from Eli Lilly Co. introduced an extra copy of *cefEF-cefG* fragment into *A. chrysogenum* which resulted in a 15%-40% higher producing mutant strain [27]. This was the first evidence that molecular breeding could be a powerful tool in strain improvement of *A. chrysogenum*.

Although controlled by the same bi-directional promoter, the transcription levels of *cefEF* and *cefG* showed a huge difference as shown by RT-PCR. The transcription of *cefG* is much lower than that of *cefEF*. This leads to the accumulation of DAC in the metabolites since they can not be efficiently transformed into CPC. As a matter of fact, CPC/DAC ratio is a quality control parameter in the industrial production of CPC fermentation. Thus, the introduction of extra copy numbers of *cefG* produced an engineering strain whose CPC yield is 3 folds higher than the parental strain [28].

There is another report on the introduction of *cefT* into *A. chrysogenum*,where the resulting mutant doubled the CPC yield [29]. This could be attributed to the enhancement of CefT, the efflux pump protein, so that the feedback inhibition *in vivo* triggered by the fermentation product was attenuated, resulting in a higher product yield.

Using molecular breeding technology, some CPC derivatives can be directly produced by engineering *A. chrysogenum* fermentation. For example, by disruption of *cefEF* and introduction of *cefE* originating from *Streptomyces clavuligerus*, a novel DAOC producing strain was obtained, which if followed by two enzymatic transformations, the industrially important 7-ADCA can be produced [30]. By introduction of the coding genes simultaneously into *A. chrysogenum* for the two enzymes used the industrial production of 7-ACA by immobilized enzymatic transformation, the engineering strain can produce 7-ACA by fermentation [31].

Besides the introduction of exogenous genes, disruption and/or silencing of the endogenous genes is also a common strategy for genetic breeding of a certain strain. The recently developed RNA interference (RNAi) technique can be used as an alternative to silence the transcription of target genes instead of homologous recombination. RNAi in *A. chrysogenum* was first published in 2007 [32]. The latest report was silencing of *pcbC* gene in *Penicillium chrysogenum* and *cefEF* gene in *A. chrysogenum* by RNAi [33]. These reports demonstrated the feasibility of RNAi technique in Filamentous fungi.

There is another interesting research for the molecular breeding of *A. chrysogenum* in a different idea. As we mentioned before, CefD1-CefD2 is a two-component enzyme complex that transforms IPN into penicillin N by an epimerization system located in the peroxisome. *cefD1-cefD2* block mutant lacking this epimerization system accumulated a large amount of IPN to more than 650μg/mL, almost the total relative CPC yield. With this mutant, the unstable IPN, which has never been purified before, could be now be purified by several steps using chromatography [34]. Characterization of its half-life and stability under a variety conditions can greatly help in the investigation of IPN.

It is worth noting that all of the above genetic breeding reports were on the background of an *A. chrysogenum* type strain C10, whose CPC yield is only 1 mg/mL, far less than the industrial production level. Although some good achievements were obtained in improvement of *A. chrysogenum* fermentation and modification of metabolic products, those achievements are still far away from application in industry.

5. Industrialization research on molecular breeding of *A. chrysogenum*

Our research is focused on the molecular breeding of *A. chrysogenum* high-yield and/or industrial strains. We introduced different combinations of *cefG*/*cefEF*/*cefT*/*vgb* genes into CPC high-producing strain and found that an extra copy of *cefG* has a significant positive effect on CPC fermentation level. Since random integration occurring in *A. chrysogenum*, different transformants with *cefG* introduction showed different elevated levels, with some at 100%. An extra copy of *vgb* gene also displayed a significant improvement up to 30% more of the CPC yield. Meanwhile, introduction of *cefEF* and *cefT* has no obvious effect on CPC production in the high-yield strain [35]. This revealed the apparent discrepancy between the genetic background of the type strain and the high-yield strain, and also suggested that endogenous *cefEF* and *cefT* may already achieve high bioactivity after several rounds of mutagenesis breeding that a high-yield strain usually undertaken.

We then applied this achievement to a CPC industrial strain. Although we didn't obtain a mutant that doubled the CPC yield, we did obtain an engineering strain whose CPC yield was increased by 20%, which has a promising industrialized potential.

We also tried the RNAi technique in the high-yield strain. A plasmid vector containing *cefG* double strain RNA transcription unit was constructed and transformed into high-yield *A. chrysogenum*. The *cefG* transcription level in the transformants was measured by quantitative RT-PCR. Two mutant strains were found to have a decreasing *cefG* transcription level of up to 80%. Their CPC yield was also found to decrease by 34.6% and 28.8%, respectively [36]. This result demonstrated the feasibility of RNAi application in high-yield *A. chrysogenum* and possible, industrial strain. Moreover, this is important for metabolic pathway reconstitution and novel CPC derivatives fermentation in *A. chrysogenum*.

The fermentation product of *A. chrysogenum*, CPC, is the major resource for industrial manufacturing of 7-ACA, the important intermediate of a large variety of cephalosporins antibiotics. A common producing route of 7-ACA is the chemical semi-biosynthesis. To date, the more environmental-friendly biotransformation has been widely used in industry. Although two step transformation dominates in the market [37], research on one step transformation from CPC to 7-ACA is still hot. However, the substrate specificity of CPC acylase still remains unsolved [38].

Whether two-step or one-step, fermentation of CPC is the prerequisite followed by enzymatic biotransformation *in vitro*. We are thinking of introducing CPC acylase gene into *A. chrysogenum* to construct the engineering strain that can produce 7-ACA directly by fermentation, a breakthrough in the production of 7-ACA.

A CPC acylase gene was designed according to the codon bias of *A. chrysogenum* and introduced into an industrial strain. Our result showed that this CPC acylase was expressed in *A. chrysogenum* with bioactivity. The recombinant acylase can transform the original product CPC into 7-ACA *in vivo*, makes the engineering strain capable of direct fermentation of 7-ACA. Based on enzymological profiles of CPC acylase *in vitro*, we performed a preliminary optimization of medium composition and culture condition and the CPC yield was increased significantly with as least 30% of the CPC fermented being transformed into 7-ACA [40]. We believe this *in vivo* conversion can be more effective if a more powerful transcription cassette and more copy number can be introduced, with the incorporation of traditional breeding technology, and finally, bring this technique to industry.

6. Perspectives

As a novel tool for strain improvement, genome shuffling is of widespread concern in the field of industrial microbiology since it was first reported [41]. This has been applied in Bacteria and Streptomyces, and the yield of a lot of metabolites got a substantial increase by genome shuffling. However, genome shuffling in Filamentous fungi is rare, maybe due to the undeveloped genetic manipulation system. In 2009, the cellulase production in *Penicillin decumbens* was reported to be increased by 40% with the help of genome shuffling [42]. But this achievement resulted largely in primary metabolites. As we all know, the regulation of secondary metabolites, as well as the genetic manipulation of *A. chrysogenum* is much more complicated. Since the exogenous genes were randomly integrated in the chromosome of *A. chrysogenum*, we suggest that genome shuffling can effectively improve the fermentation of the strains based on the established genetic techniques in our laboratory.

The biosynthesis of CPC in *A. chrysogenum* has been investigated thoroughly. However, the mechanism of its regulation as well as the biosynthesis of precursors in primary metabolism is still unclear [43]. The full sequence of *A. chrysogenum* is yet to be completed, although there are more than 10 species belonging to the Filamentous fungi that have already been sequenced [44]. To better understand the genetic basis of *A. chrysogenum*, we realize that comparative proteomics could be used to study the molecular breeding without the genomic sequence of *A. chrysogenum*. By identifying those different expressed proteins during CPC fermentation, fermentation may be proposed based on the popular theory of metabolic engineering and system biology [45].

Besides its use in studying the mechanism of disease development, the application of comparative proteomics in antibiotic-producing microorganisms also showed promise. For instance, 345 different proteins were identified as critical during the conversion from primary to secondary metabolism in *Streptomyces coelicolor* [46]. Another example is research on *Penicillin chrysogenum* where 950 proteins involved in precursor biosynthesis, stress response and pentose phosphate pathway were found to be related to the fermentation yield in 3 penicillin-producing strains [47].

Thus, we believe that the molecular breeding of *A. chrysogenum* should consist of genome shuffling, optimization of secondary metabolism, improvement of precursor biosynthesis and energy metabolism as well. Although there are still big effects need to be put in the basic and practical research of *A. chrysogenum,* the molecular bred engineering strains will play an important role in the industrial production of CPC and its derivatives.

Author details

Youjia Hu

Address all correspondence to: bebydou@hotmail.com

Shanghai Institute of Pharmaceutical Industry, Department of Biopharmaceuticals, Shanghai, China

References

[1] Ellaiah P, Adinarayana K, Chand GM, Subramanyam GS, Srinivasulu B. Strain improvement studies for cephalosporin C production by *Cephalosporium acremonium. Pharmazie,* 2002, 57(7): 489–490.

[2] Ellaiah P, Kumar JP, Saisha V, Sumitra JJ, Vaishali P. Strain improvement studies on production of cephalosporin C from *Acremonium chrysogenum* ATCC 48272. *Hindustan Antibiot Bull,* 2003, 45-46(1-4): 11–5.

[3] Lee MS, Lim JS, Kim CH, Oh KK, Yang DR, Kim SW. Enhancement of cephalosporin C production by cultivation of *Cephalosporium acremonium* M25 using a mixture of inocula. *Lett Appl Microbiol,* 2001. 32(6): 402–406.

[4] Gutierrez S, Diez B, Montenegro E, Martin JF. Characterization of the *Cephalosporium acremonium pcbAB* gene encoding alpha-aminoadipyl-cysteinyl-valine synthetase, a large multidomain peptide synthetase: linkage to the *pcbC* gene as a cluster of early cephalosporin biosynthetic genes and evidence of multiple functional domains. *J Bacteriol,* 1991, 173(7): 2354–2365.

[5] Martin JF, Ullan RV, Casqueiro J. Novel genes involved in cephalosporin biosynthesis: the three-component isopenicillin N epimerase system. *Adv Biochem Eng Biotechnol*, 2004, 88: 91–109.

[6] Gutierrez S, Velasco J, Fernandez FJ, Martin JF. The *cefG* gene of *Cephalosporium acremonium* is linked to the *cefEF* gene and encodes a deacetylcephalosporin C acetyltransferase closely related to homoserine O-acetyltransferase. *J Bacteriol*, 1992, 174(9): 3056–3064.

[7] Lejon S, Ellis J, Valegard K. The last step in cephalosporin C formation revealed: crystal structures of deacetylcephalosporin C acetyltransferase from *Acremonium chrysogenum* in complexes with reaction intermediates. *J Mol Biol*, 2008, 377(3): 935–944.

[8] Brakhage AA, Thon M, Sprote P, Scharf DH, Al-Abdallah Q, Wolke SM, Hortschansky P. Aspects on evolution of fungal beta-lactam biosynthesis gene clusters and recruitment of trans-acting factors. *Phytochemistry*, 2009, 70(15-16): 1801–1811.

[9] Dreyer J, Eichhorn H, Friedlin E, Kurnsteiner H, Kuck U. A homologue of the *Aspergillus velvet* gene regulates both cephalosporin C biosynthesis and hyphal fragmentation in *Acremonium chrysogenum*. *Appl Environ Microbiol*, 2007, 73(10): 3412–3422.

[10] Ullan RV, Teijeira F, Guerra SM, Vaca I, Martin JF. Characterization of a novel peroxisome membrane protein essential for conversion of isopenicillin N into cephalosporin C. *Biochem J*, 2010, 432(2): 227–236.

[11] Teijeira F, Ullan RV, Fernandez-Lafuente R, Martin JF. CefR modulates transporters of beta-lactam intermediates preventing the loss of penicillins to the broth and increases cephalosporin production in *Acremonium chrysogenum*. *Metab Eng*, 2011, 13(5): 532-543.

[12] Teijeira F, Ullan RV, Guerra SM, Garcia-Estrada C, Vaca I, Martin JF. The transporter CefM involved in translocation of biosynthetic intermediates is essential for cephalosporin production. *Biochem J*, 2009, 418(1): 113–124.

[13] Kuck U, Hoff B. New tools for the genetic manipulation of filamentous fungi. *Appl Microbiol Biotechnol*, 2010, 86(1): 51–62.

[14] Meyer, V. Genetic engineering of filamentous fungi--progress, obstacles and future trends. *Biotechnol Adv*, 2008, 26(2): 177–185.

[15] Skatrud PL, Queener SW, Carr LG, Fisher DL. Efficient integrative transformation of *Cephalosporium acremonium*. *Curr Genet*, 1987, 12(5): 337–348.

[16] Groot MJ, Bundock P, Hooykaas PJ, Beijersbergen AG. *Agrobacterium tumefaciens*-mediated transformation of filamentous fungi. *Nat Biotechnol*. 1998, 16: 839-842

[17] Xu W, Zhu C, Zhu B. An efficient and stable method for the transformation of heterogeneous genes into *Cephalosporium acremonium* mediated by *Agrobacterium tumefaciens. J Microbiol Biotechnol*, 2005, 15(4): 683–688.

[18] Yuan N, Hu Y, Zhu C, Zhu B. DNA shuffling of *Vitreoscilla* Hemoglobin. *China Biotechnology*, 2006, 26(11):14-19.

[19] Xu W, Zhu C, Zhu B, Yao X. A new method for isolation of chromosomal DNA from filamentous fungus *Cephalosporium acremonium. Journal of Shenyang Pharmaceutical University*. 2004, 21(3): 226–229.

[20] Zhang P, Zhu C, Zhu B, Zhao W. A convenient method to select DNA fragments of *Cephalosporium acremonium* with promoter function. *Microbiology*. 2004, 31(3): 97–100.

[21] Zhang P, Zhu C, Zhu B. Cloning of bidirectional *pcbAB-pcbC* promoter region from *Cephalosporium acremonium* and its application. *Acta Microbiologica Sinica*. 2004. 44(2): 255–257.

[22] Chen D, Yuan N, Hu Y, Zhu C, Zhao W, Zhu B. Cloning, expression and activity analysis of DAC-acetyltransferase gene from *Acremonium chrysogenum. Chinese Journal of Antibiotics*, 2006. 31(7): 395–399.

[23] Chen D, Hu Y, Zhu C, Zhu B. Optimization on soluble expression of a recombinant DAC-acetyltransferase from *Acremonium chrysogenum* and its enzyme kinetics. *Chinese Journal of Pharmaceuticals*, 2007, 38(9): 625–628.

[24] Nijland JG, Kovalchuk A, van den Berg MA, Bovenberg RA, Driessen AJ. Expression of the transporter encoded by the *cefT* gene of *Acremonium chrysogenum* increases cephalosporin production in *Penicillium chrysogenum. Fungal Genet Biol*, 2008, 45(10): 1415–1421.

[25] Lin YH, Li YF, Huang MC, Tsai YC. Intracellular expression of *Vitreoscilla* hemoglobin in *Aspergillus terreus* to alleviate the effect of a short break in aeration during culture. *Biotechnol Lett*, 2004, 26(13): 1067–1072.

[26] DeModena JA, Gutierrez S, Velasco J, Fernandez FJ, Fachini RA, Galazzo JL, Hughes DE, Martin JF. The production of cephalosporin C by *Acremonium chrysogenum* is improved by the intracellular expression of a bacterial hemoglobin. *Bio/Technology*, 1993. 11(8): 926–929.

[27] Skatrud P, Tietz A, Ingolia T, Cantwell C, Fisher D, Chapman J, Queener S. Use of recombinant DNA to improve production of cephalosporin C by *Cephalosporium acremonium. Bio/Technology*, 1989, 7(5): 477–485.

[28] Gutierrez S, Velasco J, Marcos AT, Fernandez FJ, Fierro F, Barredo JL, Diez B, Martin JF. Expression of the *cefG* gene is limiting for cephalosporin biosynthesis in *Acremonium chrysogenum. Appl Microbiol Biotechnol*, 1997, 48(5): 606–614.

[29] Ullan RV, Liu G, Casqueiro J, Gutierrez S, Banuelos O, Martin JF. The *cefT* gene of *Acremonium chrysogenum* C10 encodes a putative multidrug efflux pump protein that

significantly increases cephalosporin C production. *Mol Genet Genomics*, 2002, 267(5): 673–683.

[30] Velasco J, Luis Adrio J, Angel Moreno M, Diez B, Soler G, Barredo JL. Environmentally safe production of 7-aminodeacetoxycephalosporanic acid (7-ADCA) using recombinant strains of *Acremonium chrysogenum*. *Nat Biotechnol*, 2000, 18(8): 857–861.

[31] Isogai T, Fukagawa M, Aramori I, Iwami M, Kojo H, Ono T, Ueda Y, Kohsaka M, Imanaka H. Construction of a 7-aminocephalosporanic acid (7ACA) biosynthetic operon and direct production of 7ACA in *Acremonium chrysogenum*. *Biotechnology (N Y)*, 1991, 9(2): 188–191.

[32] Janus D, Hoff B, Hofmann E, Kuck U. An efficient fungal RNA-silencing system using the DsRed reporter gene. *Appl Environ Microbiol*, 2007, 73(3): 962–970.

[33] Ullan RV, Godio RP, Teijeira F, Vaca I, Garcia-Estrada C, Feltrer R, Kosalkova K, Martin JF. RNA-silencing in *Penicillium chrysogenum* and *Acremonium chrysogenum*: validation studies using beta-lactam genes expression. *J Microbiol Methods*, 2008, 75(2): 209–218.

[34] Vaca I, Casqueiro J, Ullan RV, Rumbero A, Chavez R, Martin JF. A preparative method for the purification of isopenicillin N from genetically blocked *Acremonium chrysogenum* strain TD189: studies on the degradation kenetics and storage conditions. *J Antibiot*, 2011, 64(6): 447-451.

[35] Liu Y, Gong G, Xie L, Yuan N, Zhu C, Zhu B, Hu Y. Improvement of cephalosporin C production by recombinant DNA integration in *Acremonium chrysogenum*. *Mol Biotechnol*, 2010, 44(2): 101–109.

[36] Gong G, Liu Y, Hu Y, Zhu C, Zhu B. Down-regulation of *cefG* gene transcription in an industrial strain of *Acremonium chrysogenum* by RNA interference. *Biotechnology Bulletin*. 2010, 10: 193–197.

[37] Conlon HD, Baqai J, Baker K, Shen YQ, Wong BL, Noiles R, Rausch CW. Two-step immobilized enzyme conversion of cephalosporin C to 7-aminocephalosporanic acid. *Biotechnol Bioeng*, 1995. 46(6): 510–513.

[38] Sonawane VC. Enzymatic modifications of cephalosporins by cephalosporin acylase and other enzymes. *Crit Rev Biotechnol*, 2006, 26(2): 95–120.

[39] Liu Y, Gong G, Hu Y, Zhu C, Zhu B. Expression of cephalosporin C acylase in *Acremonium chrysogenum*. *Chinese Journal of Pharmaceuticals*, 2009, 40(12): 902-906.

[40] Liu Y, Gong G, Zhu C, Zhu B, Hu Y. Environmentally safe production of 7-ACA by recombinant *Acremonium chrysogenum*. *Curr Microbiol*, 2010, 61(6): 609–614.

[41] Zhang YX, Perry K, Vinci VA, Powell K, Stemmer WP, del Cardayre SB. Genome shuffling leads to rapid phenotypic improvement in bacteria. *Nature*, 2002, 415(6872): 644–646.

[42] Cheng Y, Song X, Qin Y, Qu Y. Genome shuffling improves production of cellulase by *Penicillium decumbens* JU-A10. *J Appl Microbiol*, 2009, 107(6): 1837–1846.

[43] Schmitt EK, Hoff B, Kuck U. Regulation of cephalosporin biosynthesis. *Adv Biochem Eng Biotechnol*, 2004, 88: 1–43.

[44] Jones MG. The first filamentous fungal genome sequences: *Aspergillus* leads the way for essential everyday resources or dusty museum specimens? *Microbiology*, 2007, 153(Pt 1): 1–6.

[45] Thykaer J, Nielsen J. Metabolic engineering of beta-lactam production. *Metab Eng*, 2003, 5(1): 56–69.

[46] Manteca A, Sanchez J, Jung HR, Schwammle V, Jensen ON. Quantitative proteomics analysis of *Streptomyces coelicolor* development demonstrates that onset of secondary metabolism coincides with hypha differentiation. *Mol Cell Proteomics*, 2010, 9(7): 1423–1436.

[47] Jami MS, Barreiro C, Garcia-Estrada C, Martin JF. Proteome analysis of the penicillin producer *Penicillium chrysogenum*: characterization of protein changes during the industrial strain improvement. *Mol Cell Proteomics*, 2010, 9(6): 1182–1198.

Section 2

Gene Targeting in Plants

Chapter 2

Strategies for Generating Marker-Free Transgenic Plants

Borys Chong-Pérez and Geert Angenon

Additional information is available at the end of the chapter

http://dx.doi.org/10.5772/55573

1. Introduction

1.1. Why marker-free transgenic plants?

Selectable marker genes (SMGs), such as antibiotic or herbicide resistance genes, are used in nearly every plant transformation protocol to efficiently distinguish transformed from non-transformed cells. However, once a transgenic event has been selected, marker genes are generally of no use. On the contrary, the continued presence of marker genes in transgenic plants may raise public and regulatory concerns and may have technological disadvantages.

The main perceived risk is horizontal gene transfer of antibiotic resistance genes to pathogenic organisms or the transfer of herbicide resistance genes to weeds. Regulatory agencies may thus advice or require the absence of certain marker genes in commercialized transgenic plants [1].

Fears concerning SMGs center around the presence of antibiotic resistance genes in transgenic crops or its products that might reduce the efficacy of a clinically important antibiotic. A lot of attention has been spent on risk assessment concerning the transfer of antibiotic resistance genes from genetically modified (GM) plants to soil- and plant-related micro-organisms by horizontal gene transfer. For example, the transformation of bacteria in the food chain where free DNA persists in some materials for weeks, and moreover, some bacteria develop natural/chemical competence to take up DNA from the environment. In addition, in the gastrointestinal tract of humans and farm animals, DNA may remain stable for some time, particularly in the colon. However, degradation already begins before the DNA or the material containing the DNA arrives at the critical sites for horizontal gene transfer, which are generally believed to be the lower part of the small intestine, caecum, and the colon. In the case that DNA can arrive to this part, it will be mostly fractionated in pieces smaller than a gene sequence. Thus, breakdown of DNA in the gut, combined with the breakdown of the DNA due to food processing, strongly reduces the risk of dissemination [2]. Moreover, the antibiotic resistance genes that are commonly used as selectable marker genes in transgenic plants actually have a

INTECH
open science | open minds

© 2013 Chong-Pérez and Angenon; licensee InTech. This is a paper distributed under the terms of the Creative Commons Attribution License (http://creativecommons.org/licenses/by/3.0), which permits unrestricted use, distribution, and reproduction in any medium, provided the original work is properly cited.

bacterial origin [3]. Indeed bacteria have developed very sophisticated mechanisms to eliminate competitors and guarantee their own survival producing antibiotics and genes to confer resistance to these antibiotics. Thus, the contribution of horizontal transfer of antibiotic resistance genes between transgenic plants and microorganism is most likely insignificant compared to the existing exchange of such genes between bacteria [3-6].

On the other hand the escape of herbicide resistance genes to wild relatives is also a concern. Many crops are sexually compatible with wild and/or weedy relatives, then if the plants grow close one to another, crop-to-weed or crop-to-wild relative gene flow could result (reviewed by [6, 7]. The success of the introgression of a transgene in a wild relative has many barriers. Firstly, both have to grow in close proximity; secondly, both have to be flowering in overlapping time frames; thirdly, the progeny must be sufficiently fertile to propagate; and fourthly, a selective pressure should be applied (herbicide) [8]. There will only be a selective advantage for the wild relative if the herbicide is used in the habitat where the relative grows. For example, it is well known that cultivated rice is sexually compatible with perennial wild red rice (*Oryza rufipogon* Griff.), considered a harmful weed. It grows in many of the same regions, often has overlapping flowering times, and thus is a prime candidate for gene flow with cultivated rice. Indeed, Chen et al. [9] showed that the gene flow rate was 0.01% under natural conditions. This and other studies showed the risk of the transfer of transgene(s) to the wild relative or weeds. Thus precautions should be taken into account to prevent gene flow and introgression. A possible way consists in containing transgenic pollen by growing barrier crops in adjacent areas or by alternating transgenic cultivars carrying different herbicide resistance genes [10]. Other strategies consists in the creation of biological containment, to limit the transfer of pollen to plants in the surrounding area, e.g. by engineering male sterility or by delaying and/or decreasing flowering [11, 12]. Alternatively, complete removal of the marker gene should alleviate concerns regarding effects on human health and the environment.

In some specific cases, selectable marker genes are needed after selection, for example in propagation of lines with nuclear male sterility [13]. However, generally SMGs are not needed after the selection of the transgene event. On the contrary, their presence may have some technological drawbacks. It has been reported that some genes (selectable markers included) may induce pleiotropic effects under certain conditions [14, 15]. In fact, a transcriptome analysis of three *Arabidopsis* transgenic lines containing pCAMBIA3300 vector (*35S-bar-35S*) showed that they differ from their WT counterparts by expression of 7, 18 and 32 genes respectively. However, only four genes were found to be significantly different in all three lines compared with the wild type plant in glufosinate untreated plants [14]. Thereafter, 81 genes were found to be differentially expressed in the presence of glufosinate in transgenic plants, in contrast to the 3762 differentially expressed genes in WT plants. From these 81 genes 29 were specific to transgenic plants [14]. These results suggested to the authors that glufosinate or a metabolic derivative of glufosinate activates unique detoxification pathways to offset any effects on plant growth and development. Nevertheless, in the above mentioned work, no indication or study of the position effect and/or effect of transgene regulatory sequences was reported. Indeed the regulatory sequences (promoters and terminators) can influence the activity of some genes in the same T-DNA or even endogenous genes that are close to the

http://dx.doi.org/10.5772/55573

insertion site [16-18]. Furthermore, in systems where the number of efficient SMGs is limited, the re-transformation with the same SMG is precluded by its presence. This is problematic as most transformation protocols are indeed based on one or a few selectable marker genes only. Miki and McHugh [3] reported that more than 90 % of the scientific publications that use transgenic plants were based on three selection systems: the antibiotics kanamycin or hygromycin and the herbicide phosphinothricin. These outcomes provide an extra motivation to remove SMGs and other unnecessary sequences as soon as possible after selection of transgenic plants.

2. Strategies to obtain marker free transgenic plants

2.1. Transformation without selection

The most straightforward method to obtain marker-free plants is to transform without any selectable marker gene. However, most of the transformation protocols described are inefficient and just few cells integrate the foreign DNA. Nonetheless, some groups have studied the feasibility to obtain transgenic plants omitting selection. De Buck et al. [19] failed to obtain any transgenic plants when *Arabidopsis* roots were transformed via *A. tumefaciens* and shoots regenerated on non-selective media. However, in tobacco protoplast transformation, these authors obtained a total transformation frequency of 18%. Transformation protocols have important influence on these and other results. For example, in a study where *Arabidopsis* was transformed by the floral dip protocol and seedlings were grown on non-selective media, transgenic plants could be obtained with an efficiency of 3.5% [20]. In citrus, 35 plants out of 620 analyzed were transgenic in the absence of selection [21]. The main objective of the experiments mentioned until here was not to obtain marker-free plants, but they showed the possibility or not to do so. Other experiments have as a goal to obtain marker free transgenic plants. For example, in wheat transformation via micro-projectile bombardment, 23 out of 191 regenerated plants in non-selective media were transgenic (12%) [22]. Also in potato and cassava transformation *via A. tumefaciens,* without selection pressure, resulted in transformed shoots at an efficiency of 1–5% of the harvested shoots [23]. In this case the presence of chimeric plants was less than 2% of transgenic plants. Other authors that mention the possibility to obtain chimeric plants were Doshi et al.[24], which obtained a transformation frequency of 0.93% and 1.55% in triticale and wheat, respectively, without selection. These authors suggest circumventing the chimera problem with two embryogenesis cycles, where the plantlets can be regenerated from secondary embryos formed from transformed sectors within a primary somatic embryo. A report of a non-selection approach for tobacco transformation showed a transformation efficiency of 2.2-2.8% for the most effective binary vector; the authors found that the number of chimeric plants was 28-56%, which is expected taking into account the regeneration system applied [25]. In another interesting report the direct production of marker-free citrus plants under non-selective conditions was assessed [26]. In two genotypes evaluated, only one produced transgenic plants with an efficiency of 1.7%. Remarkably, the expression of the gene of interest (*sgfp*) was very low in transgenic plants. This phenomenon has been reported before in citrus [21], *Arabidopsis thaliana* [20, 27] and white pine (*Pinus strobus* L.) [28] transformed with *A. tumefaciens* and regenerated without selection. These results are in line with the fact that T-DNA integrates randomly with respect to particular DNA sequences in the genome, and that target sites include transcriptionally 'silent' regions, like telomeres [29]. Cells with insertion events in such silent regions likely do not survive when selection is applied.

2.2. Co-transformation of a marker gene and the gene-of-interest followed by segregation and selection of marker free progeny plants

Many approaches have been reported to remove selectable marker genes since the transformation technology was developed in the 80s. One of the earliest methods was based on co-transformation of a transgene and a selectable marker delivered by two separate DNA molecules and thereafter, segregation of both in the progeny (reviewed in [3, 30-32]). This strategy is based on the fact that cells selected for the presence of the marker gene, often contain the non-selected gene of interest as well. The SMG and the gene of interest can be delivered by: (i) two different *Agrobacterium* strains each containing a binary plasmid carrying a single T-DNA region (Fig. 1A) [33-37]; (ii) a single *Agrobacterium* strain, either containing one plasmid with two separate T-DNAs (Fig. 1C) [33, 36, 38-41] or (iii) containing two separate plasmids each containing a T-DNA (Fig. 1B) [42,43]. Alternatively, co-transformation can be achieved by particle co-bombardment (Fig. 1D) [44, 45]. The co-transformation strategy is limited because co-integration of both T-DNAs at the same genomic locus is frequently observed leading to linkage between the marker and the transgene, which makes their segregation impossible. This phenomenon has even more frequently been observed with particle bombardment-mediated transformation. Moreover, these methods cannot be applied to sterile plants and vegetatively propagated species, and are not practical in plants with a long life cycle such as trees [46]. This approach requires the generation of many transformants (to find unlinked marker genes and genes-of-interest) and further crossing steps (to remove the marker gene) making it a labor intensive work.

2.3. Placing the selectable marker gene or the gene-of-interest on a transposable element

Transposable elements (e.g. *Ac/Ds* from maize) can mediate repositioning of genetic material in the plant genome. The *Ac/Ds* transposable element system has been used for relocation and elimination of a selectable marker in tomato [47, 48] and rice [49]. Transposable elements can be excised from the genome after the expression of the transposase; they can either re-insert or not (Fig. 2). Taking into account these options, two approaches can be followed. In the first one, if one counts on re-insertion of the transposon, the gene of interest (GOI) is placed on the transposable element. Thus, the GOI will be excised and can be reinserted in a locus that is not linked to the locus in which the selectable marker gene is located; they can be segregated in the next generation [47, 49]. In a second approach, one relies on the fact that the transposon will not be re-inserted [50]. An example of such a system is the one described by Ebinuma et al. [51], in which the *ipt* selectable marker gene was inserted in an *Ac* derivative. However, marker-free transgenic plants were obtained only with a very low efficiency (5%) as a result of a high rate of re-insertion.

http://dx.doi.org/10.5772/55573

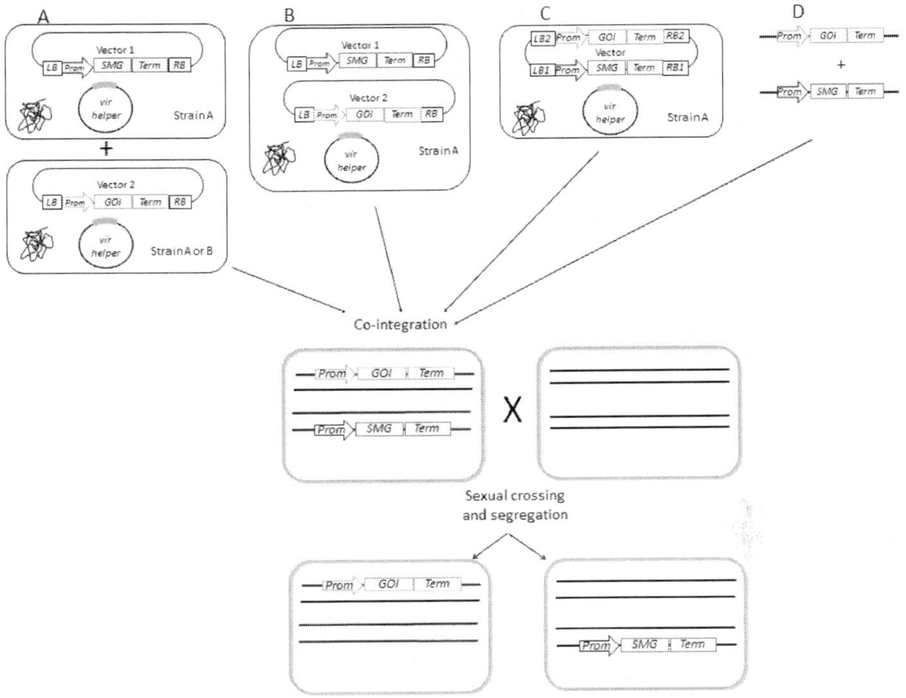

Figure 1. Co-transformation / segregation strategy to obtain marker-free transgenic plants. The SMG and the gene of interest (GOI) are introduced on separate T-DNAs present in two different *Agrobacterium strains* (A), on separate vectors in the same *Agrobacterium* strain (B), or on the same vector (C); the two genes can also be delivered by a direct gene transfer method such as particle bombardment (D). If the GOI and the SMG are integrated at unlinked positions, progeny plants with only the GOI can be obtained after sexual propagation. LB: T-DNA left border, RB: T-DNA right border, Prom: promoter; Term: terminator

This system has some advantages associated with the relocation of the gene of interest. For example, it permits to study a large range of position effects thereby generating an extensive qualitative and quantitative variation in expression levels from a single transpositionally active transformant line [49]. Moreover, relocation allows elimination through recombination in the progeny of all sequences co-integrated at the original integration site. Thus the integration pattern is simplified and the relocated transposon-borne transgene may be less susceptible to gene silencing than at the original integration [52].

On the other hand, this system has several drawbacks. First, the transposition efficiency is variable in different species. Second, the method is labor intensive and time consuming because it requires crossing transgenic plants and the selection of the progeny [53, 54]. The method shows low efficiency of marker gene elimination because of the tendency of transposable elements to reinsert in positions genetically linked to the original position. Other disadvantages of this system are the genomic instability of transgenic plants because of the continuous presence of heterologous transposons and the generation of mutations because of insertion and excision cycles. Transposition can induce genome rearrangements, including deletions, inverted duplications, inversions, and translocations [55]. Additionally, this system cannot be used for sterile plants and vegetatively propagated species and is impractical for plants with a long life cycle.

2.4. Homologous recombination

Another method developed for marker gene removal takes advantage of the DNA repair machinery of plant cells. Indeed, efficient repair of double-strand breaks (DSBs) is important for survival of all organisms. DSBs can be repaired via homologous recombination (HR) or non-homologous end-joining (NHEJ) [56]. The ratio of HR to NHEJ events increases if homologous sequences near the brake are available [57]. During the repair process the gene can be converted or deleted [58]. Orel et al. [56] showed that deletion-associated pathway was about five times more frequent than the pathway resulting in gene conversion. These findings were exploited by Zubko et al. [59], who placed the selectable marker genes between two directly repeated 352 bp *attP* regions of bacteriophage λ. This sequence is rich in A+T nucleotides that is supposed to have a stimulatory effect on recombination [60]. Moreover, these elements were situated adjacent to a copy of the transformation booster sequence (TBS) from *Petunia hybrida*, which was shown to increase both HR and NHEJ in *Petunia, Nicotiana* and maize [61]. After selection on antibiotic (kanamycin) containing media, tobacco callus was

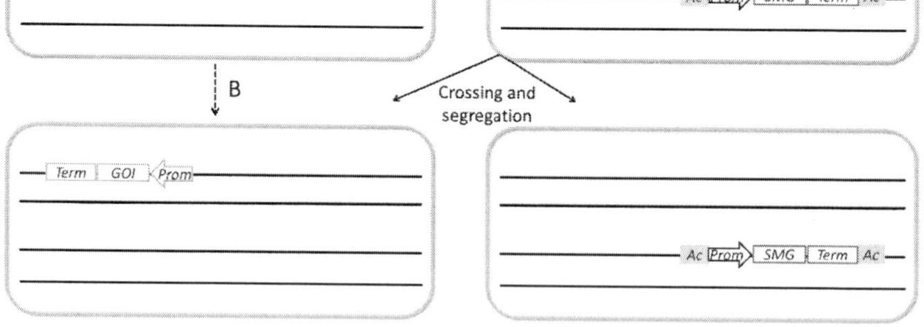

Figure 2. Transposon-mediated repositioning of the SMG. The SMG is cloned as part of a modified transposable element, e.g. the maize transposable element *Ac*, and linked to the gene of interest (GOI). Transposition may result in re-insertion of the modified element with the SMG (A); if the re-insertion occurs in an unlinked position, marker-free progeny may be obtained after crossing. Alternatively, no re-insertion occurs after excision of the modified transposable element (B), also resulting in the loss of the SMG.

placed on antibiotic-free media to allow for the loss of SMG by homologous recombination. Thereafter, plants were regenerated from callus and selection of marker free plants was based on sensitivity to the antibiotic. Two clones showed sensitivity to the antibiotic and formed green and white shoots. From these clones the authors regenerated 23 marker free plants. However, from these marker-free plants, 20 lost the gene of interest that was outside of the *attP* sites, probably because the NHEJ mechanism [59]. This protocol should produce marker-free plants faster than do procedures involving re-transformation or cross-pollination, and also avoid potential problems related with expression of recombinases (discussed below). Nonetheless, the method has some major disadvantages, like low efficiency, deletions of non-target genes, the recombination cannot be controlled and many transgenic events can be lost during the selection process. The mechanistic basis of the phenomenon is not yet understood and it is not yet known how the system could be applied in other crops.

2.5. Removal of the selectable marker gene after the selection procedure via site-specific recombinases or zinc finger nucleases

Another system to remove selectable marker genes is based on site-specific recombinases and was first reported about 20 years ago [62, 63]. Microbial site-specific recombinases have the ability to cleave DNA at specific sites and ligate it to the cleaved DNA at a second target sequence. The excision of foreign DNA that is placed in between recognition sites in a direct repeat orientation has been used to eliminate unwanted transgenic material from the nuclear genome of plants (Fig. 3). The most used recombination systems are Cre/*lox* from bacteriophage P1 [64, 65], FLP/*FRT* from *Saccharomyces cerevisiae* [66, 67] and R/*RS* from *Zygosaccharomyces rouxii* [68]. These systems are belonging to the tyrosine recombinase family [69, 70]. After the reaction, a recombination site (*lox, FRT* or *RS*) is remaining in the genome and it could potentially serve as a site for integrative recombination. However, re-insertion of the elimination fragment has not been detected [53, 71], probably because excision is an intramolecular event, whereas integration needs interaction between unlinked sites; and second, the excised circle cannot replicate autonomously and is probably rapidly lost *in vivo* [30].

The site-specific recombination systems can be divided in two categories according to the position of the recombinase gene. In a first category of strategies, the recombinase gene and the selectable marker are on a different vector and the recombinase gene is delivered to the plant containing the SMG by re-transformation [62, 72, 73] or by sexual crosses [63, 74-77].

A main limitation of both systems is that they require a time-consuming and labor-intensive breeding step, and that they are only applicable to sexually reproducing species or some species where the retransformation is available. An alternative approach depends on the expression of the recombinase transiently [78]. Marker-free plants were also obtained after infection of PPT resistant *Nicotiana benthamiana* and *Arabidopsis thaliana* leaves with a modified plant virus carrying the *cre* gene (PVX-Cre) [79-81]; as well as in kanamycin resistant tobacco with a TMV-Cre [82]. This method can be applied to vegetatively propagated and long life cycle plants, but the lack of virus-based transformation system in these species is a drawback that should be improved in the future.

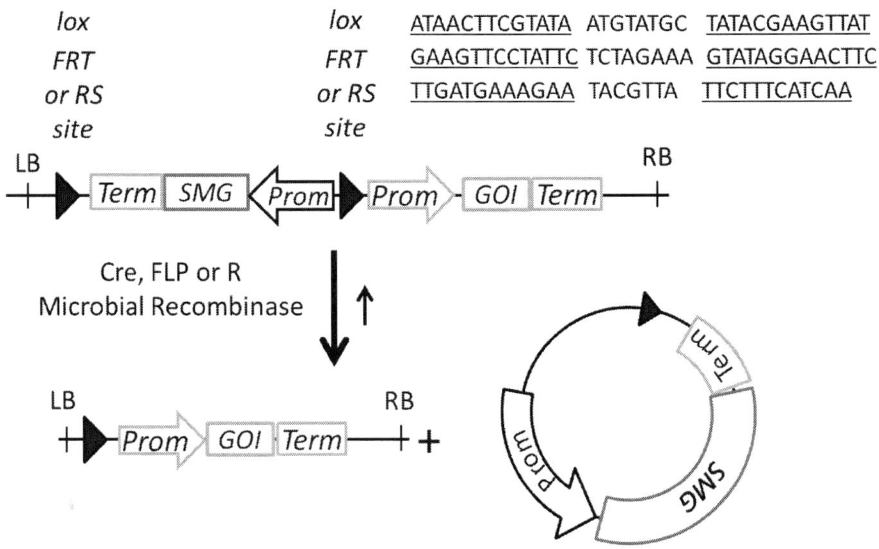

Figure 3. Removal of selectable marker genes through site specific recombinases. The SMG is flanked by directly repeated recombinase recognition sites, most often the *lox*, *FRT* or *RS* sites (black triangles; sequence of the sites is shown in the upper right corner). The presence of the cognate recombinase enzyme, Cre, FLP or R respectively, directs excision of the SMG. Re-insertion of the SMG occurs with low frequency if at all.

Nevertheless, as all technologies also the site-specific recombination systems have some drawbacks. *In vitro* studies suggest that Cre can catalyze recombination between certain naturally occurring "pseudo-*lox* sites" that can be highly divergent from the *lox* consensus sequence [83]. It was also shown that constitutive expression of *cre* can lead in animal cells to growth-inhibitory and genotoxic effects as a result of the endonuclease activity of Cre [84, 85]. This toxic effect was also investigated in *cre* expressing transgenic plants where a correlation was found between aberrant phenotypes and constitutive *cre* expression [86]. Data regarding the presence of cryptic *FRT* or *RS* sites or the infidelity of FLP- or R-recombinase activities in higher eukaryotes do not appear to be available [30]. These findings suggest it may be useful to limit *cre* expression both temporally and spatially, by placing it under the control of regulated promoters.

In a second category of methods using site-specific recombination, the selectable marker and the recombinase genes are on the same vector between the recombination sites (Fig. 4). This system is often referred to as "auto-excision" [87] or self-excision [88]. The auto-excision strategy is a versatile system that could be applied in every species and that shows flexibility in spatial and temporal control. The expression of the recombinase gene can be induced by either external or intrinsic signals resulting in auto-excision of both the recombinase and marker genes placed within the excision site boundaries after their function is no longer needed. The control of excision is enabled by the regulated promoter used to control the recombinase gene. This approach was described with heat-shock inducible promoter-recombinase expression cassettes in *Arabidopsis* [89, 90], tobacco [91, 92], potato [93], maize [94], Chinese white poplar (*Populus tomentosa* Carr.) [95], hybrid aspen (*Populus tremula* L. × *P. tremuloides* Michx.) [96] and rice [97, 98]. In the latter experiment, the selectable marker gene and *cre* gene were co-bombarded, but probably the efficiency could be higher if both were on the same vector [98]. We have recently obtained transgenic banana plants devoid of the marker gene using a Cre-*lox* auto-excision strategy induced by heat shock [99].

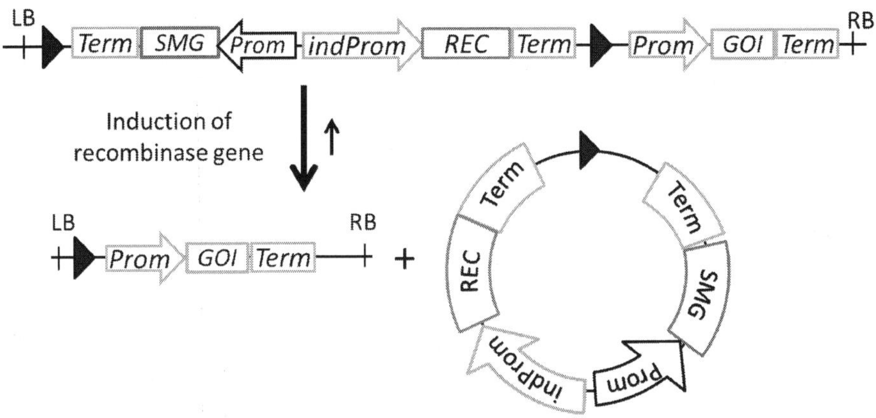

Figure 4. Site-specific recombinase based auto-excision systems. The site specific recombinase gene (*REC*) is under control of an inducible promoter (*indProm*) and is placed together with the SMG between directly repeated recombinase recognition sites (black triangles). Induction of recombinase gene expression leads to excision of the SMG and the recombinase gene.

The recombinase can also be driven by chemically regulated promoters, like the GST-II-27 promoter from maize which is induced by an herbicide antidote Safener, to control the *R/RS* system in tobacco [100] and aspen [46, 101], with β-estradiol trans-induction of *Cre/lox* in *Arabidopsis* [102], rice [103], and tomato [104] and with a dexamethasone-glucocorticoid receptor ligand binding domain activated *R/RS* system in strawberry [105] and potato [106]. In the latter cases, the authors used a combined positive–negative selection scheme to obtain marker- and recombinase-free genotypes [105, 106].

A more refined approach comprises self-excision controlled by an endogenous stimulus that is a part of the plant life cycle. For example, Mlynárová et al. [107] reported the use of a microspore-specific NTM19 promoter from tobacco to drive the expression of the *cre* gene. Thus the excision of the marker gene is taking place during the microsporogenesis where the efficiency was close to 100% in tobacco seeds. An improvement of this system was reported by Luo et al. [108] and was called 'GM-gene-deletor'. In this, the excision target unit is flanked with two different fused target sites (*lox-FRT*), as an alternative to the use of one recombination site (either *lox* or *FRT*) at each side. Activation of either recombinase (Cre or FLP) by a pollen or pollen- and seed-specific promoter PAB5, gave up to 100% excision efficiency of *lox-FRT* fusion-bounded transgenes in some transformation events, leaving residual LB and RB elements flanking a *lox-FRT* site, in both pollen and/or seed. The use of germline-specific promoters derived from the *Arabidopsis APETALA1* and *SOLO DANCERS* genes, and combined with a positive-negative selection strategy, allowed Verweire et al. [87] to produce completely marker- and recombinase-free *Arabidopsis* plants. Similarly, the expression of Cre driven by the rice floral specific *OsMADS45* gene promoter, excised the *nptII* gene flanked by *lox* recombination sites in T1 rice generation [109]. In another approach, Li et al. [110] took advantage of the somatic embryogenesis developmental stage required in soybean transformation. In this report, the activation of *cre* gene was driven by the *Arabidopsis app1* embryo-specific gene promoter and successfully directed the production of marker- and recombinase-free soybean; in 13% of the events complete excision was noted, whereas 31% yielded chimeras and in 56% of the events the excision failed [110]. Excision systems have also been developed based on seed inducible promoters. Indeed, the cruciferin C promoter from *Arabidopsis* was used to control the expression of *cre* gene in tobacco seeds [88] but the excision efficiency was low (10.2%). Additionally, in a similar strategy *Brassica napus* and tobacco marker free plants were obtained when *cre* gene was driven by a seed-specific-napin promoter form *Brassica napus* [111, 112]. In *B. napus* the efficiency ranged from 13 to 81% and in tobacco from 55 to 100% [111, 112].

The auto-excision strategy is very flexible in timing enabling the excision to take place in late (e.g. flowering or seedling) or early (e.g. somatic embryos) developmental stages. In addition many of these approaches are applicable to vegetatively propagated plant species and and long life cycle plants like perennial trees.

An additional feature of recombination based systems is the capability to resolve complex insertion sites containing multiple tandem insertions of the T-DNA down to more simple or even single copy structures. In wheat, maize and Arabidopsis it was demonstrated that complex integration patterns can be resolved by Cre-mediated recombination, thereby generating single copy transformants [87, 113, 114].

Additionally, the apparent disadvantage of the remaining presence of one *lox* site after the excision is, in some cases, an important advantage. Indeed, the marker-free transgenic line containing one *lox* site can be used as a target line for gene stacking. It has been reported that the Cre/*lox* system can be used to introduce DNA via site-specific integration [115-120]. This has three major advantages. First, the stacked trait is integrated in the genome in a genomic locus giving predictable transgene expression. Second, the trait is introduced in a locus which is already approved by the regulating authorities. Third, by stacking the traits in this way they are linked to one another facilitating breeding programs [121].

A number of novel recombinase systems have been identified that also show the ability to excise DNA in eukaryotic cells [90, 122-126]. So far, only ParA [90] and ΦC31 [126] have been effectively used in plant.

An alternative to the site-specific recombination system would be to construct restriction endonucleases that recognize specific T-DNA sequences. Zinc finger nucleases (ZFNs) could for example be used to eliminate selectable marker genes or other unnecessary DNA sequences from the plant genome. ZFNs are artificial restriction enzymes that consist of a synthetic C_2H_2 zinc finger DNA-binding domain fused to the DNA cleavage domain of the restriction enzyme *Fok*I [127, 128]. ZFNs are capable of inducing targeted double strand breaks[129]. Until now, a unique approach for ZFN-mediated transgene deletion was reported by Petolino et al. [130]. These authors crossed ZFN-overexpressing plants with target transgenic plants, which were engineered to carry a GUS expression cassette that was flanked by recognition sites for the ZFN. Both types of transgenic plants were homozygous for the transgene. The higher frequency of GUS negative hybrid plants was 35 % for one particular cross. PCR and sequencing analyses confirmed that the GUS cassette had indeed been removed.

However, many ZFNs have been reported to be toxic [131-133] presumably as a result of the creation of non-target DSBs [134]. Thus, the strategy to address this problem would be the regulation of ZFN expression by the use of inducible promoters or the use of transient expression systems like the plant virus systems mentioned above. Another approach could be the redesigning of the *Fok*I cleavage domain to create obligate heterodimers [134].

2.6. Removal of transplastome marker gene

In the last decade plastid genome (plastome or ptDNA) has become a popular target for engineering, as this has several advantages like potentially high level protein expression, maternal inheritance and non-dissemination of transgenes through pollen, high transgene copy number per cell and no detected gene silencing [135]. However, selectable marker genes are unnecessary once transplastomic plant has been obtained. Moreover high levels of marker gene expression can cause metabolic problems. Additionally, for selection only spectinomycin and streptomycin (*aadA*) or kanamycin (*nptII* or *kan* and *aphA-6*) genes have been used. Then, four strategies to produce marker-free transplastomic plant have been developed: (i) homology-based excision via directly repeated sequences, (ii) excision by phage site-specific recombinanses, (iii) transient cointegration of the marker gene, and (iv) co-transformation-segregation approach.

2.6.1. *Homology based SMG excision via directly repeated sequences*

This approach is based on the efficient native homologous recombination apparatus of the plastid. This system relies on the presence of directly repeated identical sequences of plastid DNA. Then, any sequence between them could be excised [136, 137]. The first indication of this phenomenon was observed in the unicellular alga, *Chlamydomonas reinhardtii* where homologous recombination between two direct repeats allowed marker removal under non-selective growth conditions [136]. Later experiments demonstrate marker excision in tobacco chloroplasts after transformation with a construct carrying three transgenes (*uidA, aadA* and *bar* genes) [135]. In the transformation vector, the authors placed two of the three genes under the same promoter (*Prrn* promoter of the rRNA operon, and all the genes with the same transcription terminator (*TpsbA*) (Fig. 5). Initial heteroplastomic clones were obtained by selection for spectinomycin and streptomycin resistance conferred by *aadA*. Thereafter, herbicide-resistant and -sensitive derivatives were identified in the absence of antibiotic selection [135]. Finally the recombination between repeated sequences rendered two types of stable marker-free plants: recombination R1 via the *Prrn* repeat produced herbicide resistant clones and by recombination R2 via *TpsbA,* Gus expressing clones were obtained (Fig. 5). Neither type of marker-free transgenic plants has repeated sequences. As herbicide resistance genes could not be used to directly select plastid transformants [138], this strategy is very useful to obtain marker-free highly herbicide resistant plants. Actually there are two versions of this strategy. The first one allowed visual tracking of the SMG excision by creation of a pigment-deficient zone due to the loss of a plastid photosynthetic gene *rbc*L [139]. The authors placed the *uidA* gene under control of the *PatpB* promoter. The recombination between this sequence with the native *PatpB,* which is located closed to the *rbc*L gene, allowed the deletion of a large DNA segment that comprises the *uidA-aadA-rbc*L genes. The cells lacking *rbc*L could be visually identified by their pale green color; these cells also lack the *uidA* and *aadA* genes. This approach could facilitate advanced studies that require the isolation of double mutants in distant plastid genes and the replacement of the deleted locus with site-directed mutant alleles and is not easily achieved using other methods [139].

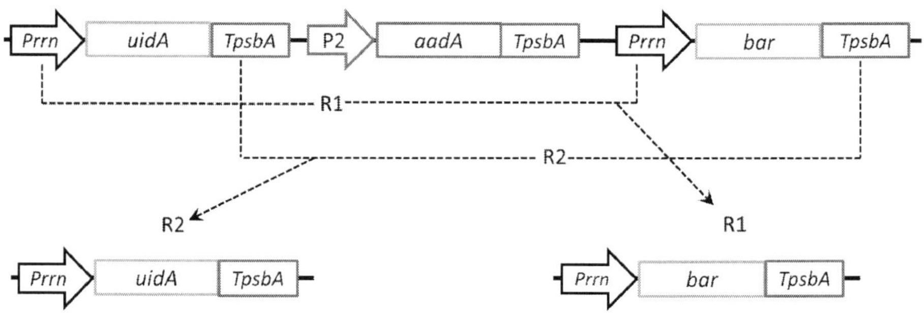

Figure 5. Homology-based marker gene excision via directly repeated sequences [135]. The repeats were the promoters (*Prrn*) and transcription terminators (*TpsbA*). Recombination via the *Prrn* promoter or *TpsbA* repeats yielded the two stable marker-free ptDNA carrying only the *uidA* (recombination R2) or only the *bar* (recombination R1) gene. No sequence is repeated in the final product. *uidA*: reporter gene encoding β-glucuronidase; *aadA*: spectinomycin resistance marker gene; *bar*: herbicide resistance gene.

In the second version [140] marker-free tobacco plants were generated by the use of a vector that harboured an *aadA* gene disrupting the herbicide resistance gene *hppd* (*4-hydroxyphenylpyruvate dioxygenase* from *Pseudomonas fluorescens* (HPPD) enzyme that confers resistance to sulcotrione and isoxaflutole). Initially, antibiotic-resistant clones were obtained. Marker-free herbicide-resistant plants were identified after excision of the *aadA* marker gene by homologous recombination within the overlapping region (403 bp) of the 5′ and 3′ halves of the herbicide resistance gene. Excision of *aadA* led to reconstitution of a complete herbicide resistance gene and expression of the HPPD (Fig. 6).

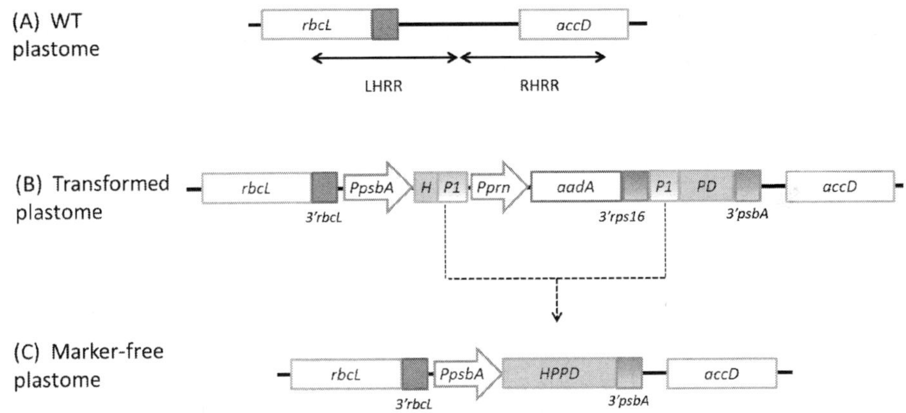

Figure 6. Homology-based marker gene excision via directly repeated sequences [140]. Integration of transgenes into the wild-type (WT) tobacco plastid genome (A) after transformation with the designed vector, giving a transformed plastome (B). After the recombination between the two P1 repeats, a marker-free plastome was obtained (C). HP1: 5′ fragment of the *4-hydroxyphenylpyruvate dioxygenase* gene (*hppd*) coding region; LHRR and RHRR, left and right homologous recombination regions; P1, repeat segment overlapping the 5′ and 3′ fragments; P1PD, 3′ fragment of the *hppd* coding region.

2.6.2. Excision by phage site-specific recombinases

Site-specific recombinases have also been used to produce marker-free transplastomic plants. This approach exploits a two-step protocol. Step one is the production of transplastomic plants, which carry a SMG flanked by two directly oriented recombinase target sites (Fig. 7). Afterward, marker-free plants could be obtained when the recombinase activity is introduced by nuclear transformation of a gene encoding a plastid-targeted recombinase[141, 142].

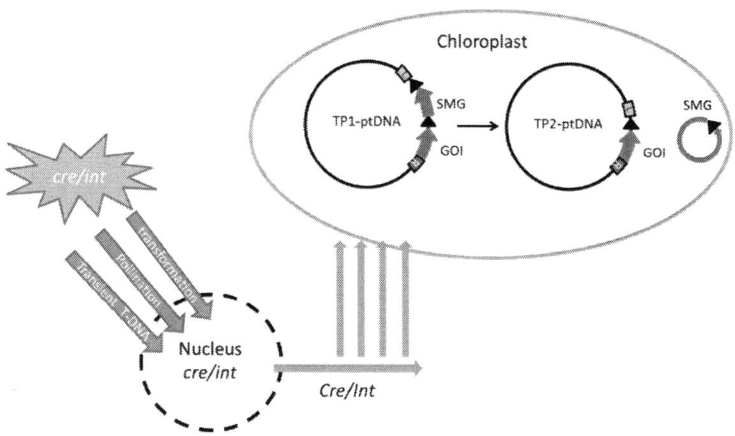

Figure 7. Marker gene excision from the plastid genome by Cre or Int site-specific recombinases [137]. A site-specific recombinase gene (*cre/int*) introduced into the nucleus by transformation, pollination or transient Agroinfiltration, encodes a plastid-targeted recombinase that excises selectable marker gene (SMG) from TP1-ptDNA after import into plastids. Excision of the marker gene by phage recombinases via the target sites (black triangles) yields marker-free TP2-ptDNA carrying only the gene of interest (GOI) and one recombinase recognition sequence [141-143].

Cre/*lox* was the first site-specific recombination system used to excise the SMG from the plastid genome [141, 142]. In these works Cre activity was introduced by nuclear transformation and marker-free transplastomic plants were obtained in tissue culture. However, these plants still contain *cre* and *nptII* genes in their genome that had to be segregated away in the seed progeny [141, 142]. Another way to introduce Cre activity is by pollination, in which apparently non-specific Cre-induced re-arrangements between homologous ptDNA sequences were absent or occurred significantly less often than in directly transformed plants [141]. On the other hand, Lutz et al. [144] took into account the fact that not every T-DNA delivery results in stable integration and expressed Cre transiently from T-DNA introduced by Agroinfiltration. As a result in this experiment approximately 10% of the regenerated plants did not carry either a plastid selectable marker or a nuclear *cre* gene. Nevertheless, Cre-mediated marker excision can cause the deletion of ptDNA sequences by recombination via directly repeated non-*lox* sequences that result in mutation of target plant [141, 142].

As an alternative, the ΦC31 phage site-specific integrase (Int) that mediates recombination between bacterial (*attB*) and phage (*attP*) attachment sites was tested to excise the SMG [143]. The authors tested marker gene excision in a two-step process. In the first step, tobacco chloroplast were transformed with a vector that contains the SMG (*aadA* gene) flanked with directly oriented non-identical phage *attP* (215 bp) and bacterial *attB* (54 bp) recombination sites, which are recognised by Int recombinase. The *bar* gene was used as gene of interest and it was placed outside of the excision cassette. Spectinomycin-resistant clones were obtained and these were stable in the absence of Int. In the second step, a plastid-targeted Int was introduced by *Agrobacterium*-mediated nuclear transformation that directed efficient marker gene excision. No fortuitous sequences appear to be present in the plastid genome that would be recognized by the Int [143]. In the homology-mediated marker excision the frequency of deletion is proportional to the length of the repeat. As the *lox* sequences are short (34 bp in length), the probability to cause loss of the marker gene in the absence of Cre is not completely absent. However the *attB* and *attP* sequences are not homologous, therefore plastid genomes carrying *att*-flanked marker genes are predicted to be more stable than those with marker genes flanked by identical *lox* sequences. The absence of homology between the *attB* and *attP* sites and the absence of pseudo-*att* sites in ptDNA would make Int a preferred alternative to Cre for plastid marker excision [137].

2.6.3. Transient cointegration of the marker gene

Based on the mechanism of integration of the foreign DNA in the plastid genome, Klaus et al. [145] designed a system to excise the SMG. Indeed, two homologous recombination events (Left and Rigth) are needed for DNA integration. However, considering that cointegrate formation is a common phenomenon that takes place in bacterial plasmid recombination, the authors assumed that in the chloroplast a transformation vector first forms a cointegrate following recombination between a single region of homology in the transformation vector and the plastome (Fig. 8A). Cointegrates are naturally unstable due to the presence of direct repeats in these molecules. Subsequent homology recombination events (between duplicated sequences) lead either to stable integration of both the GOI and SMG gene or to loss of the integrated vector, yielding a wild-type plastome (Fig. 8A) [145]. In this work the authors used a vector where the marker gene (*aphA-6*) was located outside of the recombination region. This strategy allowed the selection for a cointegrate structure that forms by recombination via only one of the targeting sequences (Fig. 8B). When selection for kanamycin resistance was withdrawn, the second recombination event can take place and the marker gene is lost.

2.6.4. Co-transformation-segregation

The co-transformation-segregation method in plastid transformation technology is based on the same principle that has been applied in nuclear transformation. Indeed, the SMG and the gene of interest are inserted in two different plasmids and introduced into two locations (Fig. 9A) of the same plastid by biolistic transformation to generate heteroplastomic cells with both or either of the genes (Fig. 9B) [137, 146]. After segregation, a marker-free transplastomic plant could be obtained (Fig. 9C). The approach was developed to obtain antibiotic resistance gene-free plants with resistance to herbicides (glyphosate or phosphinothricin) due to the impossibility to obtain such plants directly. Indeed, transplastomic plants cannot be obtained directly by selection with herbicides after transformation with the resistance genes (*CP4* or *bar*) because cells harboring only a few copies of the transgene die [147, 148]. Nonetheless, when these genes were co-transformed with a plasmid carrying the spectinomycin resistance (*aadA*) gene and most ptDNA copies carry the genes, the cells and regenerated plants showed resistance to high levels of the herbicides [147, 148].

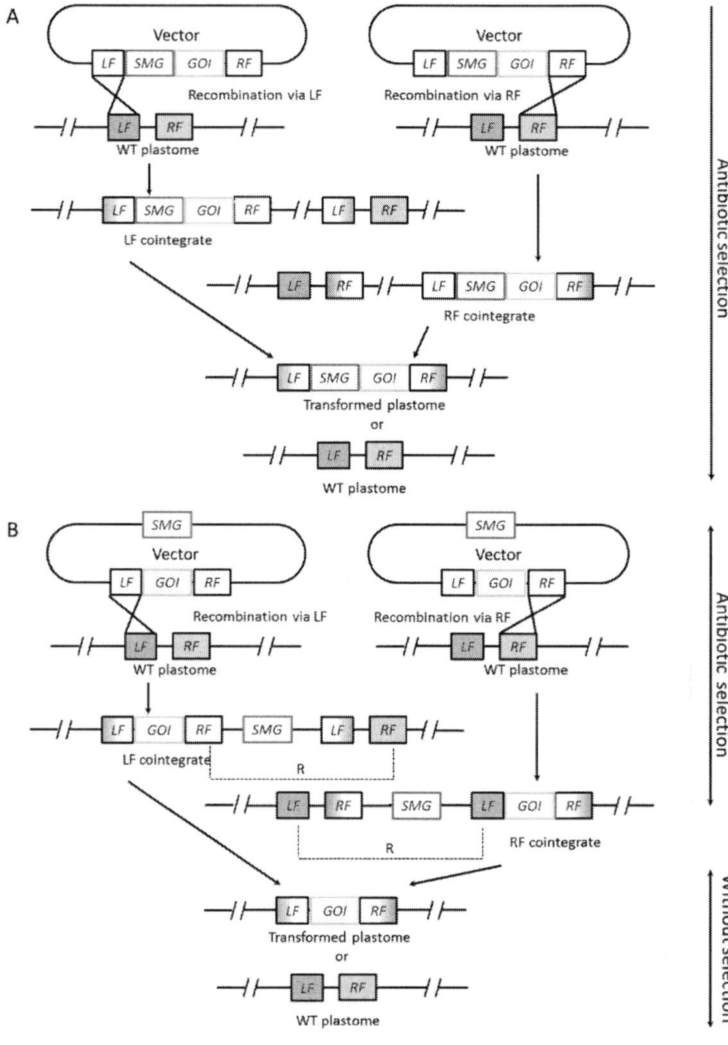

Figure 8. Transient cointegration of the marker gene to obtain marker-free transplastomic plants [145]. Cointegrate formation and subsequent recombination events with conventional and alternative plastid transformation vectors. (A) Standard plastid transformation using a vector in which the SMG is cloned between the homologous flanks. Recombination via a single flank (left or right) results in cointegration of the vector; subsequent loop-out recombination events between direct repeats lead either to a stably transformed plastome containing the sequence of interest and marker or wild-type plastome. (B) Plastid transformation using a vector in which the selection marker is cloned outside of the homologous flanks. Again recombination via either left or right flanks results in cointegration of the vector; however, following additional recombination events only the GOI is stably integrated and the marker gene is lost.

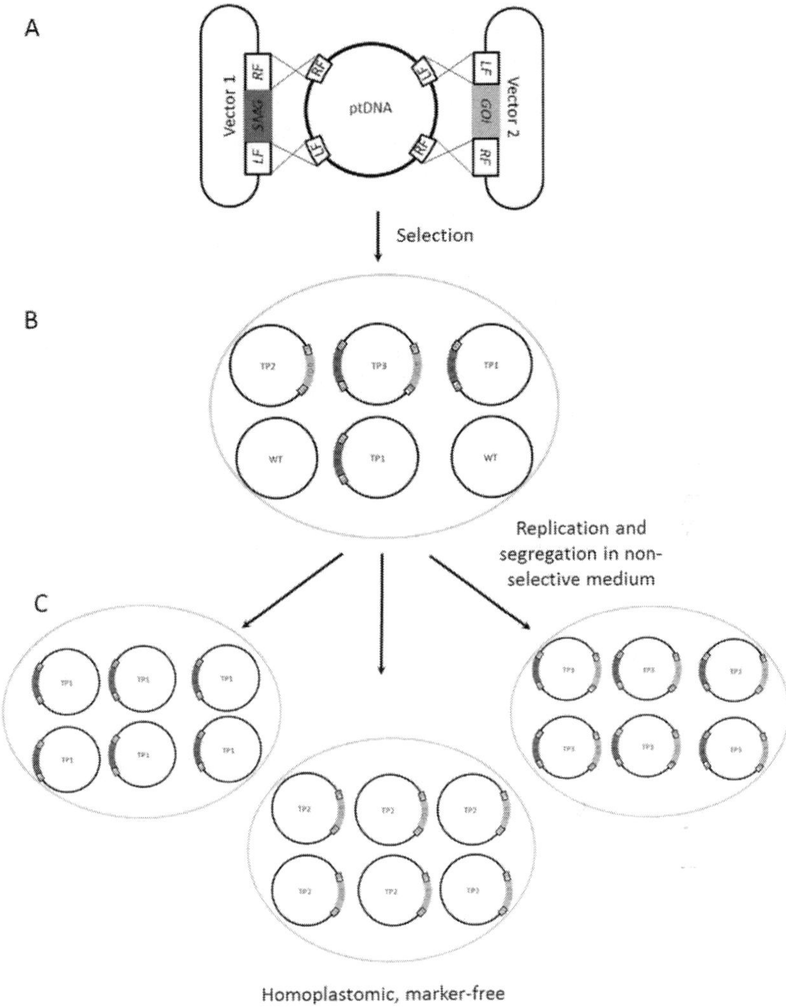

Figure 9. The cotransformation-segregation method to remove selectable marker genes from transplastomic plants [149]. (A) Transformation of the plastid genome (ptDNA) with two vectors. Vector 1 containing the selectable marker gene (SMG) and Vector 2 the gen of interest (GOI). (B) Transplastomic clones are identified by selection for antibiotic resistance. The heteroplastomic cell carries wild-type ptDNA (wt), TP1-ptDNA obtained by transformation with Vector 1; TP2-ptDNA transformed with Vector 2; and TP3-ptDNA transformed with both vectors. (C) Replication and segregation of ptDNA on non-selective medium eventually yields homoplastomic cells with TP1-ptDNA, TP2-ptDNA and TP3-ptDNA. Desired marker-free transplastomic plants carry TP2-ptDNA and lack the antibiotic resistance marker (adapted from [137]).

3. Marker free transgenic plants with agronomically useful genes

The various methods to obtain marker-free transgenic plants have proven their utility and are increasingly being deployed to obtain crop plants with agronomically useful genes. One of the crops that have received more attention is rice. Indeed some papers have described the production of marker-free transgenic plants with different genes of interest. Applying the co-transformation / segregation strategy with the use of 'double right border' twin T-DNA vectors Lu et al. [39] obtained marker-free transgenic rice plants harboring a *Rice ragged stunt virus* (RRSV) derived synthetic resistance gene. Another group obtained transgenic rice devoid of selectable marker genes that produce high levels of carotenoids using the same strategy but with two binary vectors in one *Agrobacterium* strain [150]. One of the vectors contained in the T-DNA the phytoene synthase (*psy*) and phytoene desaturase (*crt*I) expression cassettes whereas the other vector contained the *hph, nptII* and *gus* genes. Marker-free rice plants, with improved resistance to *Magnaporthe grisea* were obtained by the expression of the rice pistil-predominant chitinase gene using a vector system with two T-DNAs [151].

Sripriya et al. [152] generated marker-free transgenic plants with improved resistance to sheath blight. A single *A. tumefaciens* strain harbored a cointegrate vector with the *hph* and *gus* genes and a binary vector with the rice chitinase (*chi1*) gene. The elimination of SMG was accomplished by segregation in T_1 progeny. Some of the lines showed an enhanced resistance to *Rhizoctonia solani*. Thereafter, the same group sequentially retransformed one of the *chi1* lines with the tobacco osmotin *ap24* gene by co-transformation using an *Agrobacterium* strain harboring a single-copy cointegrate vector pGV2260::pSSJ1 (*hph* and *gus*) and a multi-copy binary vector pBin19ΔnptII-ap24 in the same cell [153]. They obtained one line in the T_1 progeny where the SMG was absent and *chi1* and *ap24* genes were integrated. Homozygous plants with both genes were obtained and some of those showed enhanced resistance to *R. solani*. Selectable marker-free rice plants expressing the *Bacillus thuringiensis* synthetic *cry1B* gene were obtained by transposon-mediated repositioning of the GOI [49]. The Cry1B expression cassette was flanked by the inverted terminal repeats of the maize *Ac* transposon that permit the repositioning of this cassette in the rice genome. Preliminary bioassays suggested that the T-DNA free relocation events exhibit a level of resistance to a major rice insect pest, *Chilo suppressalis*.

On the other hand, Sengupta et al. [154] have exploited the Cre/*lox* site-specific recombination system to produce selectable marker-free transgenic rice plants with improved resistance to green leafhoppers (*Nephotettix virescens*) and brown planthopper (*Nilaparvata lugens*). In this work two independent vectors were used, one having the *ASAL* (*Allium sativum* leaf agglutinin) gene and the *hpt* gene flanked by *lox* sites, and the other with the *cre* and *bar* genes. Cre activity was introduced by crossing single copy T_0 plants and marker excision was detected in T_1 hybrids. T_2 progeny showed the segregation of the *cre-bar* T-DNA and improved insect resistance.

The first commercially available marker-free transgenic plant that was obtained through this system was developed by the company Renessen. They generated the transgenic corn line LY038, from which the *nptII* selectable marker gene, originally present between tandemly oriented *lox* sites, was removed through introduction of the *cre* gene by a sexual cross [121]. For the market, this corn line has the name Mavera™ High Value Corn with Lysine, and was developed for the feed industry. This line was obtained from a biolistic transformation event where a *cordapA* gene encoding a seed-specifically expressed lysine insensitive dihydrodipicolinate synthase enzyme. This approach was also applied to obtain marker-free salt tolerant maize plants by the expression *AtNHX1*, a Na^+ /H^+ antiporter gene from *Arabidopsis*, but using the FLP/*FRT* system [155].

In another report, the production of marker-free transgenic soybean [*Glycine max* (L.) Merr.] is described which produces γ-linolenic acid and stearidonic acid that are important for pharmaceutical and nutraceutical industries [41]. The authors applied the co-transformation / segregation strategy, using a vector with two T-DNAs: the first harbored a cDNA of the *Borago officinalis* L. Δ^6 desaturase gene driven by the embryo- specific β-conglycinin promoter, whereas the second T-DNA contained the selectable marker gene *bar*. In this work ~7% of the transgenic lines were marker-free.

The expression of a chitinase gene, *ChiC*, on an *ipt*-type MAT (isopentenyl transferase-type multi-auto-transformation) vector, allowed the production of marker-free disease-resistant transgenic potato plants [156]. Based on transformation without selectable marker gene Stiller et al. [157] obtained two potato lines with increased levels of the semi-essential amino acid cysteine by expression of the serine acetyltransferase encoding *cys*E gene. This system was also used by Ahmad et al. [158] to produce marker-free potato plants with enhanced tolerance to oxidative stress by the expression of the superoxide dismutase and ascorbate peroxidase genes. The same strategy was used in Chinese cabbage (*Brassica campestris* ssp. *pekinensis* (Lour) Olsson) to produce Turnip mosaic virus (TuMV) resistant marker-free transgenic plants [159]. This approach was also applied in melon (*Cucumis melo*), where ripening was delayed by the introduction of marker-free and vector-free antisense 1-aminocyclopropane-1-carboxylic acid oxidase (ACC oxidase) construct [160].

Author details

Borys Chong-Pérez[1] and Geert Angenon[2*]

*Address all correspondence to: geert.angenon@vub.ac.be

1 Instituto de Biotecnología de las Plantas, Universidad Central "Marta Abreu" de Las Villas, Santa Clara, Cuba

2 Laboratory of Plant Genetics, Vrije Universiteit Brussel, Brussels, Belgium

References

[1] EFSA (2004) Opinion of the Scientific Panel on Genetically Modified Organisms on the use of antibiotic resistance genes as marker genes in genetically modified plants. EFSA Journal 48, 1-18.

[2] Van den Eede G, Aarts H, Buhk H-J, Corthier G, Flint HJ, Hammes W, Jacobsen B, Midtvedt T, van der Vossen J, von Wright A, Wackernagel W, Wilcks A (2004) The relevance of gene transfer to the safety of food and feed derived from GM plants. Food Chem. Toxicol. 42: 1127-1156

[3] Miki B, McHugh S (2004) Selectable marker genes in transgenic plants: applications, alternatives and biosafety. J. Biotechnol. 107: 193-232

[4] Thompson J (2000). Topic 11: gene transfer—mechanism and food safety risks. Joint FAO/WHO Expert Consultation on Foods Derived from Biotechnology, Geneva.

[5] Bennett PM, Livesey CT, Nathwani D, Reeves DS, Saunders JR, Wise R (2004) An assessment of the risks associated with the use of antibiotic resistance genes in genetically modified plants: report of the Workig Party of the British Society for Antimicrobial Chemotherapy. J. Antimicrob. Chemother. 53:418-431

[6] Ramessar K, Peremarti A, Gómez-Galera S, Naqvi S, Moralejo M, Muñoz P, Capell T, Christou P (2007) Biosafety and risk assessment framework for selectable marker genes in transgenic crop plants: a case of the science not supporting the politics. Transgenic Res. 16: 261-280

[7] Ellstrand NC (2003) Current knowledge of gene flow in plants: implications for transgene flow. Phil. Trans. R. Soc. Lond. B. 358: 1163-1170

[8] Mallory-Smith C, Zapiola M (2008) Gene flow from glyphosate-resistant crops. Pest Manag. Sci. 64: 428-440

[9] Chen LJ, Lee DS, Song ZP, Suh HS, Lu B-R (2004) Gene flow from cultivated rice (*Oryza sativa*) to its weedy and wild relatives. Ann. Bot. 93: 67-73

[10] Lemaux PG (2009) Genetically engineered plants and foods: a scientist's analysis of the issues (part II). Annu. Rev. Plant Biol. 60: 511-559

[11] Mascia PN, Flavell RB (2004) Safe and acceptable strategies for producing foreign molecules in plants. Curr. Opin. Plant Biol. 7: 189-195

[12] Kwit C, Moon HS, Warwick SI, Stewart CN (2011) Transgene introgression in crop relatives: molecular evidence and mitigation strategies. Trends Biotechnol. 29: 284-293

[13] Denis M, Delelourme R, Gourret J-P, Mariani C, Renard M (1993) Expression of engineered nuclear male sterility in *Brassica napus*. Plant Physiol. 101: 1295-1304

[14] Abdeen A, Miki B (2009) The pleiotropic effects of the BAR gene and glufosinate on the *Arabidopsis* transcriptome. Plant Biotechnol. J. 7: 266-282

[15] Miki B, Abdeen A, Manabe Y, MacDonald P (2009) Selectable marker genes and unintended changes to the plant transcriptome. Plant Biotechnol. J. 7: 211-218

[16] Ouakfaoui SE, Miki B (2005) The stability of the *Arabidopsis* transcriptome in transgenic plants expressing the marker genes *nptII* and *uidA*. Plant J. 41: 791-800

[17] Yoo SY, Bomblies K, Yoo SK, Yang JW, Choi MS, Lee JS, Weigel D, Ahn JH (2005) The 35S promoter used in a selectable marker gene of a plant transformation vector affects the expression of the transgene. Planta 221: 523-530

[18] Zheng XL, Deng W, Luo KM, Duan H, Chen YQ, McAvoy R, Song SQ, Pei Y, Li Y (2007) The cauliflower mosaic virus (CaMV) 35S promoter sequence alters the level and patterns of activity of adjacent tissue- and organ-specific gene promoters. Plant Cell Rep. 26: 1195-1203

[19] De Buck S, Jacobs A, Van Montagu M, Depicker A (1998) *Agrobacterium tumefaciens* transformation and cotransformation frequencies of *Arabidopsis thaliana* root explants and tobacco protoplasts. Mol. Plant-Microbe Interact. 11: 449-457

[20] Francis KE, Spiker S (2005) Identification of *Arabidopsis thaliana* transformants without selection reveals a high occurrence of silenced T-DNA integrations. Plant J. 41: 464-477

[21] Domínguez A, Fagoaga C, Navarro L, Moreno P, Peña L (2002) Regeneration of transgenic citrus plants under non selective conditions results in high-frequency recovery of plants with silenced transgenes. Mol. Genet. Genomics 267: 544-556

[22] Permingeat HR, Alvarez ML, Cervigni GD, Ravizzini RA, Vallejos RH (2003) Stable wheat transformation obtained without selectable markers. Plant Mol. Biol. 52: 415-419

[23] De Vetten N, Wolters AM, Raemakers K, van der Meer I, ter Stege R, Heeres E, Heeres P, Visser R (2003) A transformation method for obtaining marker-free plants of a cross-pollinating and vegetatively propagated crop. Nat. Biotechnol. 21: 439-442

[24] Doshi KM, Eudes F, Laroche A, Gaudet D (2007) Anthocyanin expression in marker free transgenic wheat and triticale embryos. In Vitro Cell. Dev. Biol. – Plant 43: 429-435

[25] Li B, Xie C, Qiu H (2009) Production of selectable marker-free transgenic tobacco plants using a non-selection approach: chimerism or escape, transgene inheritance, and efficiency. Plant Cell Rep. 28: 373-386

[26] Ballester A, Cervera M, Peña L (2010) Selectable marker-free transgenic orange plants recovered under non-selective conditions and through PCR analysis of all regenerants. Plant Cell Tissue Organ Cult. 102: 329-336

[27] Kim SI, Veena, Gelvin SB (2007) Genome-wide analysis of *Agrobacterium* T-DNA integration sites in the *Arabidopsis* genome generated under non-selective conditions. Plant. J. 51: 779-791

[28] Tang W, Newton RJ, Weidner DA (2007) Genetic transformation and gene silencing mediated by multiple copies of a transgene in eastern white pine. J. Exp. Bot. 58: 545-554

[29] Gelvin SB, Kim SI (2007) Effect of chromatin upon *Agrobacterium* T-DNA integration and transgene expression. Biochim. Biophys. Acta 1769: 409-420

[30] Hare PD, Chua NH (2002) Excision of selectable marker genes from transgenic plants. Nat. Biotechnol. 20: 575-580

[31] Puchta H (2003) Marker-free transgenic plants. Plant Cell Tissue Organ Cult. 74: 123-134

[32] Darbani B, Eimanifar A, Stewart CN Jr, Camargo WN (2007) Methods to produce marker-free transgenic plants. Biotechnol. J. 2: 83-90

[33] Depicker A, Herman L, Jacobs A, Schell J, Van Montagu M (1985) Frequencies of simultaneous transformation with different T-DNAs and their relevance to the *Agrobacterium*/plant cell interaction. Mol. Gen. Genet. 201: 477-484

[34] McKnight TD, Lillis MT, Simpson RB (1987) Segregation of genes transferred to one plant cell from two separate Agrobacterium strains. Plant Mol. Biol. 8: 439-445

[35] De Block M, Debrouwer D (1991). Two T-DNA's co-transformed into *Brassica napus* by a double *Agrobacterium tumefaciens* infection are mainly integrated at the same locus. Theor. Appl. Genet. 82: 257-263

[36] Komari T, Hiei Y, Saito Y, Murai N, Kumashiro T (1996) Vectors carrying two separate T-DNAs for co-transformation of higher plants mediated by *Agrobacterium tumefaciens* and segregation of transformants free from selection markers. Plant J. 10: 165-174

[37] Poirier Y, Ventre G, Nawrath C (2000) High-frequency linkage of co-expressing T-DNA in transgenic *Arabidopsis thaliana* transformed by vacuum-infiltration of *Agrobacterium tumefaciens*. Theor. Appl. Genet. 100: 487-493

[38] Xing AQ, Zhang ZY, Sato S, Staswick P, Cemente T (2000) The use of the two T-DNA binary system to derive marker-free transgenic soybeans. In Vitro Cell. Dev. Biol. – Plant 36: 456-463

[39] Lu HJ, Zhou XR, Gong ZX, Upadhyaya NM (2001) Generation of selectable marker free transgenic rice using double right-border (DRB) binary vectors. Aust. J. Plant. Physiol. 28: 241-248

[40] McCormac AC, Fowler MR, Chen DF, Elliott MC (2001) Efficient co-transformation of *Nicotiana tabacum* by two independent T-DNAs, the effect of T-DNA size and implications for genetic separation. Transgenic Res. 10: 143-155

[41] Sato S, Xing A, Ye X, Schweiger B, Kinney A, Graef G, Clemente T (2004) Production of γ-Linolenic Acid and Stearidonic Acid in Seeds of Marker-Free Transgenic Soybean. Crop Science 44: 646-652

[42] de Framond A, Back E, Chilton W, Kayes L, Chilton M-D (1986) Two unlinked T-DNAs can transform the same tobacco plant cell and segregate in the F1generation. Mol. Gen. Genet. 202: 125-131

[43] Daley M, Knauf VC, Summerfelt KR, Turner JC (1998) Co-transformation with one *Agrobacterium tumefaciens* strain containing two binary plasmids as a method for producing marker-free transgenic plants. Plant Cell Rep. 17: 489-496

[44] Zhao Y, Qian Q, Wang H-Z, Huang D-N (2007) Co-transformation of gene expression cassettes via particle bombardment to generate safe transgenic plant without any unwanted DNA. In Vitro Cell. Dev. Biol.-Plant. 43: 328-334

[45] Shiva Prakash N, Bhojaraja R, Shivbachan SK, Hari Priya GG, Nagraj TK, Prasad V, Srikanth Babu V, Jayaprakash TL, Dasgupta S, Spencer TM, Boddupalli RS (2009) Marker-free transgenic corn plant production through co-bombardment. Plant Cell Rep. 28: 1655-1668

[46] Matsunaga E, Sugita K, Ebinuma H (2002) Asexual production of selectable marker-free transgenic woody plants, vegetatively propagated species. Mol. Breed. 10: 95-106

[47] Goldsbrough AP, Lastrella CN, Yoder JI (1993) Transposition mediated re-positioning and subsequent elimination of marker gene from transgenic tomato. Biotechnology 11: 1286-1292

[48] Yoder JI, Goldsbrough AP (1994) Transformation systems for generating marker-free transgenic plants. Bio-Technology 12: 263-267

[49] Cotsaftis O, Sallaud C, Breitler JC, Meynard D, Greco R, Pereira A, Guiderdoni E (2002) Transposon-mediated generation of marker free rice plants containing a Bt endotoxin gene conferring insect resistance. Mol. Breeding 10: 165-180

[50] Gorbunova V, Levy AA (2000) Analysis of extrachromosomal Ac/Ds transposable elements. Genetics 155: 349-359

[51] Ebinuma H, Sugita K, Matsunaga E, Yamakado M (1997) Selection of marker-free transgenic plants using the iso-pentenyl transferase gene as a selectable marker. Proc. Natl. Acad. Sci. U.S.A. 94: 2117-2121

[52] Koprek T, Rangel S, McElroy D, Louwerse JD, Williams-Carrier RE, Lemaux PG (2001) Transposon-mediated single copy gene delivery leads to increased transgene expression stability in barley. Plant Physiol. 125: 1354-1362

[53] Ebinuma H, Sugita K, Matsunaga E, Endo S, Yamada K, Komamine A (2001) Systems for the removal of a selection marker and their combination with a positive marker. Plant Cell Rep. 20: 383-392

[54] Ow DW (2001) The right chemistry for marker gene removal? Nat. Biotechnol. 19: 115-116

[55] Yu C, Zhang J, Peterson T (2011) Genome rearrangements in maize induced by alternative transposition of reversed *Ac/Ds* termini. Genetics 188: 59-67

[56] Orel N, Kyryk A, Puchta H (2003) Different pathways of homologous recombination are used for the repair of double-strand breaks within tandemly arranged sequences in the plant genome. Plant J. 35: 604-612

[57] Siebert R, Puchta H (2002) Efficient repair of genomic double-strand breaks by homologous recombination between directly repeated sequences in the plant genome. Plant Cell 14: 1121-1131

[58] Fishman-Lobell J, Rudin N, Haber JE (1992) Two alternative pathways of double-strand break repair that are kinetically separable and independently modulated. Mol. Cell. Biol. 12: 1292-1303

[59] Zubko E, Scutt C, Meyer P (2000) Intrachromosomal recombination between attP regions as a tool to remove selectable marker genes from tobacco transgenes. Nat. Biotechnol. 18: 442-445

[60] Muller AE, Kamisugi Y, Gruneberg R, Niedenhof I, Horold RJ, Meyer P (1999) Palindromic sequences and A+T-rich DNA elements promote illegitimate recombination in *Nicotiana tabacum*. J. Mol. Biol. 291: 29-46

[61] Galliano H, Muller AE, Lucht JM, Meyer P (1995) The transformation booster sequence from *Petunia hybrida* is a retrotransposon derivative that binds to the nuclear scaffold. Mol. Gen. Genet. 247: 614-622

[62] Dale EC, Ow DW (1991) Gene transfer with subsequent removal of the selection gene from the host genome. Proc. Natl. Acad. Sci. U.S.A. 88: 10558-10562

[63] Russell SH, Hoopes JL, Odell JT (1992) Directed excision of a transgene from the plant genome. Mol. Gen. Genet. 234: 49-59

[64] Hoess RH, Ziese M, Sternberg N (1982) P1 site-specific recombination: nucleotide sequence of the recombining sites. Proc. Natl. Acad. Sci. U.S.A. 79: 3398-3402

[65] Hoess RH, Abremski K (1985) Mechanism of strand cleavage and exchange in the Cre-*lox* site-specific recombination system. J. Mol. Biol. 181: 351-362

[66] Cox MM (1983) The FLP protein of the yeast 2 mm plasmid: expression of a eukaryotic genetic recombination system in *Escherichia coli*. Proc. Natl. Acad. Sci. U.S.A. 80: 4223-4227

[67] Senecoff JF, Bruckner RC, Cox MM (1985) The FLP recombinase of the yeast 2-mm plasmid: characterization of its recombination site. Proc. Natl. Acad. Sci. U.S.A. 82: 7270-7274

[68] Araki H, Jearnpipatkul A, Tatsumi H, Sakurai T, Ushio K, Muta T, Oshima Y (1985) Molecular and functional organization of yeast plasmid pSR1. J. Mol. Biol. 182: 191-203

[69] Gidoni D, Srivastava V, Carmi N (2008) Site-specific excisional recombination strategies for elimination of undesirable transgenes from crop plants. In Vitro Cell. Dev. Biol. – Plant 44: 457-467

[70] Wang Y, Yau Y-Y, Perkins-Balding D, Thomson JG (2011) Recombinase technology: applications and possibilities. Plant Cell Rep. 30: 267-285

[71] Zuo J, Chua N-H (2000) Chemical-inducible systems for regulated expression of plant genes. Curr. Opin. Biotechnol. 11: 146-151

[72] Odell J, Caimi P, Sauer, B, Russell S (1990) Site-directed recombination in the genome of transgenic tobacco. Mol. Gen. Genet. 223: 369-378

[73] Lyznik LA, Rao KV, Hodges TK (1996) FLP-mediated recombination of *frt* sites in the maize genome. Nucleic Acids Res. 24: 3784-3789

[74] Bayley CC, Morgan M, Dale EC, Ow DW (1992) Exchange of gene activity in transgenic plants catalyzed by the Cre-*lox* site-specific recombination system. Plant Mol. Biol. 18: 353-361

[75] Kilby NJ, Davies GJ, Snaith MR (1995) FLP recombinase in transgenic plants: constitutive activity in stably transformed tobacco and generation of marked cell clones in Arabidopsis. Plant J. 8: 637-652

[76] Hoa TTC, Bong BB, Huq E, Hodges TK (2002) Cre/*lox* site-specific recombination controls the excision of a transgene from the rice genome. Theor. Appl. Genet. 104: 518-525

[77] Kerbach S, Lorz H, Becker D (2005) Site-specific recombination in *Zea mays*. Theor. Appl. Genet. 111: 1608-1616

[78] Gleave AP, Mitra DS, Mudge SR, Morris BAM (1999) Selectable marker-free transgenic plants without sexual crossing: transient expression of *cre* recombinase and use of a conditional lethal dominant gene. Plant Mol. Biol. 40: 223-235

[79] Kopertekh L, Juttner G, Schiemann J (2004a) PVX-Cre-mediated marker gene elimination from transgenic plants. Plant Mol. Biol. 55: 491-500

[80] Kopertekh L, Jüttner J, Schiemann J (2004b) Site-specific recombination induced in transgenic plants by PVX virus vector expressing bacteriophage P1 recombinase. Plant Sci. 166: 485-492

[81] Kopertekh L, Schiemann J (2005) Agroinfiltration as a tool for transient expression of cre recombinase *in vivo*. Transgenic Res. 14: 793-798

[82] Jia H, Pang Y, Chen X, Fang R (2006) Removal of the selectable marker gene from transgenic tobacco plants by expression of Cre recombinase from a Tobacco Mosaic Virus vector through agroinfection. Trangenic Res. 15: 375-384

[83] Thyagarajan G, Guimaraes MJ, Groth AC, Calos MP (2000) Mammalian genomes contain active recombinase recognition sites. Gene 244: 47-54

[84] Loonstra A, Vooijs M, Beverloo HB, Al Allak B, van Drunen E, Kanaar R, Berns A, Jonkers J (2001) Growth inhibition and DNA damage induced by Cre recombinase in mammalian cells. Proc. Natl. Acad. Sci. U.S.A. 98: 9209-9214

[85] Buerger A, Rozhitskaya O, Sherwood MC, Dorfman AL, Bisping E, Abel ED, Pu WT, Izumo S, Jay PY (2006) Dilated cardiomyopathy resulting from high-level myocardial expression of Cre-recombinase. J. Card. Fail. 12: 392-398

[86] Coppoolse ER, de Vroomen MJ, Roelofs D, Smit J, van Gennip F, Hersmus BJ, Nijkamp HJ, van Haaren MJ (2003) Cre recombinase expression can result in phenotypic aberrations in plants. Plant Mol. Biol. 51: 263-279

[87] Verweire D, Verleyen K, De Buck S, Claeys M, Angenon G (2007) Marker-free transgenic plants through genetically programmed auto-excision. Plant Physiol. 145: 1220-1231

[88] Moravčíková J, Vaculková E, Bauer M, Libantová, J (2008) Feasibility of the seed specific cruciferin C promoter in the self excision Cre/*lox*P strategy focused on generation of marker-free transgenic plants. Theor. Appl. Genet. 117: 1325-1334

[89] Hoff T, Schnorr K-M, Mundy J (2001) A recombinase-mediated transcriptional induction system in transgenic plants. Plant Mol. Biol. 45: 41-49

[90] Thomson JG, Yau YY, Blanvillain R, Chiniquy D, Thilmony R, Ow DW (2009) ParA resolvase catalyzes site-specific excision of DNA from the Arabidopsis genome. Transgen Res. 18: 237-248

[91] Liu HK, Yang C, Wei ZM (2005) Heat shock-regulated site-specific excision of extraneous DNA in transgenic plants. Plant Sci. 168: 997-1003

[92] Wang Y, Chen B, Hu Y, Li J, Lin Z (2005) Inducible excision of selectable marker gene from transgenic plants by the Cre/*lox* site-specific recombination system. Transgenic Res. 14: 605-614

[93] Cuellar W, Gaudin A, Solorzano D, Casas A, Nopo L, Chudalayandi P, Medrano G, Kreuze J, Ghislain M (2006) Self-excision of the antibiotic resistance gene *nptII* using a heat inducible *Cre-loxP* system from transgenic potato. Plant Mol. Biol. 62: 71-82

[94] Zhang W, Subbarao S, Addae P, Shen A, Armstrong C, Peschke V, Gilbertson L (2003) Cre/*lox*-mediated marker gene excision in transgenic maize (*Zea mays* L.) plants. Theor. Appl. Genet. 107: 1157-1168

[95] Deng W, Luo K, Li Z, Yang Y (2009) A novel method for induction of plant regeneration via somatic embryogenesis. Plant Sci. 177: 43-48

[96] Fladung M, Schenk TMH, Polak O, Becker D (2010) Elimination of marker genes and targeted integration via FLP/*FRT* recombination system from yeast in hybrid aspen (*Populus tremula* L. × *P. tremuloides* Michx.). Tree Genet. Gen. 6: 205-217

[97] Akbudak MA, Srivastava V (2011) Improved FLP recombinase, FLPe, efficiently removes marker gene from transgene locus developed by Cre–*lox* mediated site-specific gene integration in rice. Mol. Biotech. 49: 82-89

[98] Khattri A, Nandy S, Srivastava V (2011) Heat-inducible Cre-*lox* system for marker excision in transgenic rice. J. Biosci. 36: 37-42

[99] Chong-Pérez B, Kosky RG, Reyes M, Rojas L, Ocaña B, Tejeda M, Pérez B, Angenon G (2012) Heat shock induced excision of selectable marker genes in transgenic banana by the Cre-lox site-specific recombination system. J. Biotechnol. 159: 265-273

[100] Sugita K, Kasahara T, Matsunaga E, Ebinuma H (2000) A transformation vector for the production of marker-free transgenic plants containing a single copy transgene at high frequency. Plant J. 22: 461-469

[101] Ebinuma H, Komamine A (2001) MAT (multiauto-transformation) vector system. The oncogenes of *Agrobacterium* as positive markers for regeneration and selection of marker-free transgenic plants. In Vitro Cell. Dev. Biol.-Plant 37: 103-113

[102] Zuo J, Niu QW, Moller SG, Chua NH (2001) Chemical-regulated, site-specific DNA excision in transgenic plants. Nat. Biotechnol. 19: 157-161

[103] Sreekala C, Wu L, Gu K, Wang D, Tian D Yin Z (2005) Excision of a selectable marker in transgenic rice (*Oryza sativa* L.) using a chemically regulated CRE/*loxP* system. Plant Cell Rep. 24: 86-94

[104] Zhang Y, Li H, Ouyang B, Lu Y, Ye Z (2006) Chemical-induced autoexcision of selectable markers in elite tomato plants transformed with a gene conferring resistance to lepidopteran insects. Biotechnol. Lett. 28: 1247-1253

[105] Schaart JG, Krens FA, Pelgrom KTB, Mendes O, Rouwendal GJA (2004) Effective production of marker-free transgenic strawberry plants using inducible site-specific recombination and a bifunctional selectable marker gene. Plant Biotechnol. J. 2: 233-240

[106] Kondrák M, van der Meer IM, Bánfalvi Z (2006) Generation of marker- and backbone-free transgenic potatoes by site-specific recombination and a bi-functional marker gene in a non-regular one-border *Agrobacterium* transformation vector. Transgenic Res. 15: 729-737

[107] Mlynárová L, Conner AJ, Nap JP (2006) Directed microspore-specific recombination of transgenic alleles to prevent pollen-mediated transmission of transgenes. Plant Biotechnol. J. 4: 445-452

[108] Luo K, Duan H, Zhao D, Zheng X, Deng W, Chen Y, Stewart CN Jr, McAvoy R, Jiang X, Wu Y, He A, Pei Y, Li Y (2007) 'GM-gene-deletor': fused *loxP*-FRT recognition sequences dramatically improve the efficiency of FLP or CRE recombinase on transgene excision from pollen and seed of tobacco plants. Plant Biotechnol. J. 5: 263-274

[109] Bai X, Wang Q, Chengcai C (2008) Excision of a selective marker in transgenic rice using a novel Cre/*loxP* system controlled by a floral specific promoter. Transgenic Res. 17: 1035-1043

[110] Li Z, Xing A, Moon BP, Burgoyne SA, Guida AD, Liang H, Lee C, Caster CS, Barton JE, Klein TM, Falco SC (2007) A Cre/*loxP*- mediated self-activating gene excision system to produce marker gene free transgenic soybean plants. Plant Mol. Biol. 65: 329-341

[111]] Kopertekh L, Broer I, Schiemann, J (2009) Developmentally regulated site-specific marker gene excision in transgenic *B. napus* plants. Plant Cell Rep. 28: 1075-1083

[112] Kopertekh L, Schulze K, Frolov A, Strack D, Broer I, Schiemann, J (2010) Cre-mediated seed-specific transgene excision in tobacco. Plant Mol. Biol. 72: 597-605

[113] De Buck S, Peck I, De Wilde C, Marjanac G, Nolf J, De Paepe A, Depicker A (2007) Generation of single-copy T-DNA transformants in Arabidopsis by the CRE/loxP recombination-mediated resolution system. Plant Physiol. 145:1171-1182

[114] Srivastava V, Anderson OD, Ow DW (1999) Single-copy transgenic wheat generated through the resolution of complex integration patterns. Proc. Natl. Acad. Sci. USA 96: 11117-11121

[115] Vergunst AC, Hooykaas PJJ (1998) Cre/lox-mediated site-specific integration of *Agrobacterium* T-DNA in *Arabidopsis thaliana* by transient expression of *cre*. Plant Mol. Biol. 38: 393-406

[116] Vergunst AC, Jansen LET, Hooykaas PJJ (1998a) Site-specific integration of *Agrobacterium* T-DNA in *Arabidopsis thaliana* mediated by Cre recombinase. Nucleic Acids Res. 26: 2729-2734

[117] Albert H, Dale EC, Lee E, Ow DW (1995) Site-specific integration of DNA into wild-type and mutant lox sites placed in the plant genome. Plant J. 7:649-659

[118] Day CD, Lee E, Kobayashi J, Holappa LD, Albert H, Ow DW (2000) Transgene integration into the same chromosome location can produce alleles that express at a predictable level, or alleles that are differentially silenced. Genes Dev. 14:2869-2880

[119] Srivastava V, Ow DW (2002) Biolistic mediated site-specific integration in rice. Mol. Breed. 8:345–350

[120]] Srivastava V, Ariza-Nieto M, Wilson AJ (2004) Cre-mediated site-specific gene integration for consistent transgene expression in rice. Plant Biotechnol. J. 2:169-179

[121] Ow DW (2007). GM maize from site-specific recombination technology, what next? Curr. Opin. Biotechnol. 18: 115-120

[122] Kempe K, Rubtsova M, Berger C, Kumlehn J, Schollmeier C, Gils M (2010) Transgene excision from wheat chromosomes by phage phiC31 integrase. Plant Mol. Biol. 72: 673-687

[123] Thomason LC, Calendar R, Ow DW (2001) Gene insertion and replacement in *Schizosaccharomyces pombe* mediated by the *Streptomyces* bacteriophage phiC31 site-specific recombination system. Mol. Genet. Genomics 265: 1031-1038

[124] Keravala A, Groth AC, Jarrahian S, Thyagarajan B, Hoyt JJ, Kirby PJ, Calos MP (2006) A diversity of serine phage integrases mediate site-specific recombination in mammalian cells. Mol. Genet. Genomics 276: 135-146

[125] Thomson JG, Ow DW (2006) Site-specific recombination systems for the genetic manipulation of eukaryotic genomes. Genesis 44: 465-476

[126] Thomson JG, Chan R, Thilmony R, Yau Y-Y, Ow DW (2010) PhiC31 recombination system demonstrates heritable germinal transmission of site-specific excision from the Arabidopsis genome. BMC Biotechnol. 10: 17

[127] Durai S, Mani M, Kandavelou K, Wu J, Porteus MH, Chandrasegaran S (2005) Zinc finger nucleases: custom-designed molecular scissors for genome engineering of plant and mammalian cells. Nucleic Acids Res. 33: 5978-5990

[128] Porteus MH, Carroll D (2005) Gene targeting using zinc finger nucleases. Nat. Biotechnol. 23: 967-973

[129] Kim YG, Cha J, Chandrasegaran S (1996) Hybrid restriction enzymes: zinc finger fusions to *Fok*I cleavage domain. Proc. Natl. Acad. Sci. USA 93: 1156-1160

[130] Petolino JF, Worden A, Curlee K, Connell J, Strange Moynahan TL, Larsen C Russell S (2010) Zinc finger nuclease-mediated transgene deletion. Plant Mol. Biol. 73: 617-628.

[131] Bibikova M, Golic M, Golic KG, Carroll D (2002) Targeted chromosomal cleavage and mutagenesis in *Drosophila* using zinc-finger nucleases. Genetics 161: 1169-1175

[132] Porteus MH (2006) Mammalian gene targeting with designed zinc finger nucleases. Mol. Ther. 13: 438-446.

[133] Pruett-Miller SM, Reading DW, Porter SN, Porteus MH (2009) Attenuation of zinc finger nuclease toxicity by small-molecule regulation of protein levels. PLoS Genet, 5, e1000376.

[134] Tzfira T, Weinthal D, Marton I, Zeevi V, Zuker A, Vainstein A (2012) Genome modifications in plant cells by custom-made restriction enzymes. Plant Biotechnol. J. 10: 373-389

[135] Iamtham S, Day A (2000) Removal of antibiotic resistance genes from transgenic tobacco plastids. Nat. Biotechnol. 18:1172-1176.

[136] Fischer N, Stampacchia O, Redding K, Rochaix JD (1996) Selectable marker recycling in the chloroplast. Mol. Gen. Genet. 251: 373-380

[137] Lutz KA, Maliga P (2007) Construction of marker-free transplastomic plants. Curr. Opin. Biotechnol. 18: 107-114

[138] Verma D, Daniel H (2007) Chloroplast vector systems for biotechnology applications. Plant Physiol. 145: 1129-1143.

[139] Kode V, Mudd EA, Iamtham S, Day A (2006) Isolation of precise plastid deletion mutants by homology-based excision: a resource for site-directed mutagenesis, multigene changes and high-throughput plastid transformation. Plant J. 46: 901-909

[140] Dufourmantel N, Dubald M, Matringe M, Canard H, Garcon F, Job C, Kay E, Wisniewski JP, Ferullo JM, Pelissier B, Sailland A, Tissot G (2007) Generation and characterization of soybean and marker-free tobacco plastid transformants over-expressing a bacterial 4-hydroxyphenylpyruvate dioxygenase which provides strong herbicide tolerance. Plant Biotechnol. J. 5: 118-133

[141] Corneille S, Lutz K, Svab Z, Maliga P (2001) Efficient elimination of selectable marker genes from the plastid genome by the CRE-*lox* site-specific recombination system. Plant J. 27: 171-178

[142] Hajdukiewicz PTJ, Gilbertson L, Staub JM (2001) Multiple pathways for Cre/*lox*-mediated recombination in plastids. Plant J. 27:161-170

[143] Kittiwongwattana C, Lutz KA, Clark M, Maliga P (2007) Plastid marker gene excision by the phiC31 phage site-specific recombinase. Plant Mol. Biol. 64: 137-143

[144] Lutz KA, Svab Z, Maliga P (2006) Construction of marker-free transplastomic tobacco using the Cre-*loxP* site-specific recombination system. Nat. Protoc. 1: 900-910.

[145] Klaus SMJ, Huang FC, Golds TJ, Koop H-U (2004) Generation of marker-free plastid transformants using a transiently cointegrated selection gene. Nat. Biotechnol. 22: 225-229

[146] Manimaran P, Ramkumar G, Sakthivel K, Sundaram RM, Madhav MS, Balachandran (2011) Suitability of non-lethal marker and marker-free systems for development of transgenic crop plants: Present status and future prospects. Biotechnol. Advances 29: 703-714

[147] Lutz KA, Knapp JE, Maliga P (2001) Expression of bar in the plastid genome confers herbicide resistance. Plant Physiol.125: 1585-1590

[148] Ye GN, Hajdukiewicz PTJ, Broyles D, Rodriquez D, Xu CW, Nehra N, Staub JM (2001) Plastid-expressed 5-enolpyruvylshikimate-3-phosphate synthase genes provide high level glyphosate tolerance in tobacco. Plant J. 25: 261-270.

[149] Ye GN, Colburn S, Xu CW, Hajdukiewicz PTJ, Staub JM (2003) Persistance of unselected transgenic DNA during a plastid transformation and segregation approach to herbicide resistance. Plant Physiol. 133: 402-410

[150] Parkhi V, Kumar V, Sunilkumar G, Campbell LM, Singh NK, Rathore KS (2009) Expression of apoplastically secreted tobacco osmotin in cotton confers drought tolerance. Mol. Breed. 23: 625-639

[151] Hashizume F, Nakazaki T, Tsuchiya T, Matsuda T (2006) Effectiveness of genotype-based selection in the production of marker-free and genetically fixed transgenic lineages: ectopic expression of a pistil chitinase gene increases leaf-chitinase activity in transgenic rice plants without hygromycin-resistance gene. Plant Biotechnol. 23: 349-356

[152] Sripriya R, Raghupathy V, Veluthambi K (2008) Generation of selectable marker free sheath blight resistant transgenic rice plants by efficient co-transformation of a cointegrate vector T-DNA and a binary vector T-DNA in one *Agrobacterium tumefaciens* strain. Plant Cell Rep. 27: 1635-1644

[153]] Ramana Rao MVR, Parameswari C, Sripriya R, Veluthambi K (2011) Transgene stacking and marker elimination in transgenic rice by sequential *Agrobacterium*-mediated co-transformation with the same selectable marker gene. Plant Cell Rep. 30: 1241-1252

[154] Sengupta S, Chakraborti D, Mondal HA, Das S (2010) Selectable antibiotic resistance marker gene-free transgenic rice harbouring the garlic leaf lectin gene exhibits resistance to sap-sucking planthoppers. Plant Cell Rep. 29: 261-271

[155] Li B, Li N, Duan X, Wei A, Yang A, Zhang J (2010) Generation of marker-free transgenic maize with improved salt tolerance using the FLP/*FRT* recombination system. J. Biotechnol. 145: 206-213

[156] Khan RS, Nyui VO, Chin DP, Nakamura I, Mii M (2011) Production of marker-free disease-resistant potato using isopentenyl transferase gene as a positive selection marker. Plant Cell Rep. 30: 587-597

[157] Stiller I, Dancs G, Hesse H, Hoefgen R, Banfalvi Z (2007) Improving the nutritive value of tubers: Elevation of cysteine and glutathione contents in the potato cultivar White Lady by marker-free transformation. J. Biotechnol. 128: 335-343

[158] Ahmad R, Kim YH, Kim MD, Phung MN, Chung WI, Lee HS, Kwak SS, Kwon SY (2008) Development of selection marker-free transgenic potato plants with enhanced tolerance to oxidative stress. J. Plant. Biol. 51: 401-407

[159] Zhandong Y, Shuangyi Z, Qiwei H 2007 High level resistance to Turnip mosaic virus in Chinese cabbage (*Brassica campestris* ssp *pekinensis* (Lour) Olsson) transformed with the antisense Nib gene using marker-free *Agrobacterium tumefaciens* infiltration. Plant Sci. 172: 920-929

[160] Hao J, Niu Y, Yang B, Gao F, Zhang L, Wang J, Hasi A. (2011) Transformation of a marker-free and vector-free antisense ACC oxidase gene cassette into melon via the pollen-tube pathway. Biotechnol. Let. 33: 55-61

Chapter 3

Gene Targeting and Genetic Transformation of Plants

Richard Mundembe

Additional information is available at the end of the chapter

http://dx.doi.org/10.5772/56335

1. Introduction

A broad definition of gene targeting includes any method that can lead to permanent site-specific modification of the genome [1], preferably with predetermined outcomes. More specifically, gene targeting is the alteration of a specific DNA sequence in an endogenous gene at its original locus in the genome, and often refers to the conversion of the endogenous gene into a designed sequence [2]. Rapid developments in the field of gene targeting, and the potential of the technology to revolutionalise genomics and plant biotechnology in particular has led to the adoption of this broad definition, over earlier definitions such as that by [3] and [4] that restricted gene targeting to homologous recombination mechanisms.

While gene targeting does not necessarily lead to marker-free, vector backbone-free transformation, gene targeting certainly brings these desired outcomes of plant transformation research closer. Such marker-free, vector backbone-free plants will be truly and precisely engineered plants, and might actually be non-transgenic, depending on the source of the sequences used. Gene targeting in *Drosophila*, mice and yeast is now more or less routine [5]. Transgenic organisms for use in research are 'made-to-order' via gene targeting and are sold by commercial companies. Gene targeting in animals is accomplished via homologous recombination (HR). However, the same cannot be said of plants. Approaches adapted from gene targeting in yeast, insect and animal models have failed to give comparable results in plants mainly because the predominant mechanism of recombination in somatic cells of plants is not HR, but is non-homologous end joining, NHEJ, also known as illegitimate recombination [6].

Double-stranded breaks in plant genomic DNA are repaired either via HR or NHEJ [7]. Homologous recombination mechanisms involve linkage of DNA fragments to regions of identical sequence, such as the other member of the homologous partner, as template for accurate repair of the double stranded break. This mechanism is therefore only functional in the S/G2 phase of the cell cycle. Non-homologous end-joining mechanisms of recombination however are functional in all phases of the life cycle and do not require significant homology

INTECH
open science | open minds

© 2013 Mundembe; licensee InTech. This is a paper distributed under the terms of the Creative Commons Attribution License (http://creativecommons.org/licenses/by/3.0), which permits unrestricted use, distribution, and reproduction in any medium, provided the original work is properly cited.

to join two fragments of a broken DNA molecule. While HR has been very successful in insects and animals [8], it has remained unavailable for the manipulation of plant transformation; it is NHEJ that is useful for plant transformation.

Occurrence of single-stranded breaks on a DNA molecule does not normally pose a challenge to the plant genome because these can be repaired by ligation without change to the nucleotide sequence. Faithful strand replacement or nick translation may take place starting at the single-strand break, again with no changes to the nucleotide sequence.

Double-stranded breaks, however, have dire consequences if not repaired, or if repaired incorrectly. A double-stranded break effectively results in two fragments of the chromosome, and only one of the fragments might have a centromere to enable separation after cell division; the other fragment might be 'lost'. Also, if unprotected, the double-stranded breaks are exposed to the exonucleases of the cell and may be misconstrued as foreign and will therefore be degraded.

Living cells therefore need efficient mechanisms for detecting chromosomal double-stranded breaks and initiation of appropriate repair mechanisms for replication to be successful. The repair of double-stranded breaks takes place by one of two main pathways for double-stranded break repair: the HR pathway or the NHEJ pathway, or both. Coincidentally, these are the two mechanisms by which exogenous DNA may also integrate into a host genome [7].

2. Homologous and non-homologous recombination

Recombination evolved in nature to repair DNA damage that may occur during the cell cycle, and to generate diversity through meiotic recombination of genetic material which in turn has enabled sexually-reproducing eukaryotes to become extremely adaptive to their ever-changing environment and is partly responsible for their success on earth.

2.1. Homologous recombination

In homologous recombination (HR) a long and extended region of homology such as that found between sister chromatids is required for the two DNA molecules to line up adjacent to each other. There are many variations to this pathway, but the basics of two popular models are illustrated in Figure 1. A cellular protein, Spo I, may induce double-stranded breaks in the chromosome. These double-stranded breaks are repaired exclusively by HR using one of several possible homologous matrices: copied from elsewhere in the genome (ectopic HR), copied from the homologue (allelic HR), or copied from the same chromosome (intra-chromosomal HR) [6, 9].

Ectopic HR is a minor pathway, and was reported to be responsible for the repair of only one in 10 000 double-stranded breaks. In some of the cases, both homologous and non-homologous end-joining mechanisms were involved in repairing different ends of the same double-stranded break [6, 10]. Of the possible ectopic recombination models, the synthesis-dependent strand annealing (SDSA) model is the one that is conservative and is consistent with these observations. Figure 1(a) below illustrates this model.

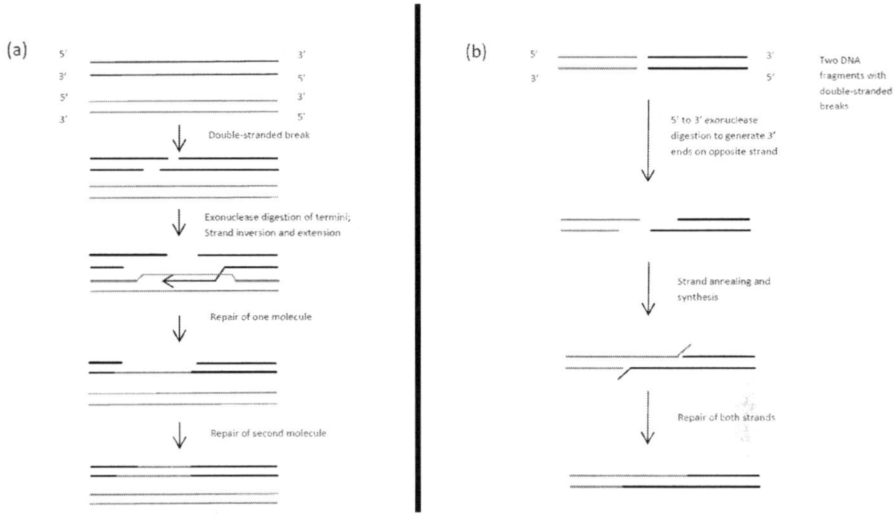

Figure 1. Models for double-stranded break repair mechanisms. (a) The synthesis-dependent strand-annealing (SDSA) for ectopic recombination. (b) The single-strand annealing (SSA) model. Polarity of the DNA molecules is shown on the first set of molecules only for simplicity. Modified from [7].

This model predicts that double-stranded break repair is not accompanied by crossing-over. Also, both perfect integrations into target sites by homologous recombination and imperfect integrations by HR are possible on one end of the target site. Integration by NHEJ is possible on the other end of the same double-stranded break of transgene, as well as ectopic integrations elsewhere in the genome, after copying of transgene sequences.

Allelic HR occurs during meiosis, to repair double-stranded breaks using sequences of the homologues in a process that involves formation of Holliday junctions to resolve into the crossover or gene conversion products. Allelic HR is not significant in somatic cells but is the classic HR that occurs in meiotic cells. In nature this essential process takes place during meiosis I to result in recombination for sexually reproducing species. Extensive lengths of homology (several hundreds or thousands of nucleotides) are required for this process, and ensures that recombination takes place between sister chromatids.

Intrachromosomal HR utilizes sequences close to the double-stranded break, on the same chromosome or on the sister chromatid (in G_2 stage only) as a matrix for repair. This can result in deletion as predicted by single-strand annealing (SSA) model (Figure 1b) or gene conversion as predicted by the conservative SDSA model depending on the structure of the chromosomal locus [11]. The SSA pathway was shown to be five time more efficient than the SDSA pathway [12]. SSA-like pathways have also been described for NHEJ.

2.2. Non-homologous end-joining recombination

The second pathway is non-homologous end-joining (NHEJ) pathway, also known as illegitimate recombination. It also requires some homology, albeit much reduced. This limited homology is required at the ends of the DNA strands on which the double-stranded breaks occur. The double-stranded breaks can be sticky ends or blunt ends. The homology present within the sticky ends may be sufficient for this mechanism, and the properly aligned ends will be ligated together. For blunt ends, binding of a specific protein complex, such as the Ku complex in mammalian cells, to the broken ends of the DNA limits nucleolytic degradation, and unlike HR repair, prevents exposure of single-stranded regions [8]. The bound protein may also function directly or indirectly to bring the DNA ends together for processing and ligation. Alignment of the termini by complementary micro-homologies of 1 – 4 nucleotides is usually required. The process might also require either limited unwinding or limited exonucleolytic digestion to expose the ends for alignment, and DNA polymerase to fill-in gaps. Single-stranded deletion of short segments at the 5′-end may expose single-stranded regions that will be used to search for homologies in the other DNA fragment, which will then form the basis of the alignment and repair [8]. The process of NHEJ is illustrated in Figure 2. The arrangement of chromosomal DNA into loops attached to a matrix that restricts the mobility of DNA promotes the re-joining of previously linked DNA ends [8].

NHEJ is the predominant pathway for double-stranded break repair in somatic cells of higher eukaryote, including plants. Simple ligation will result in junctions with no homology. Short stretches of homology may be a result of SSA-like mechanisms [13], while longer stretches might be from an SDSA copying of ectopic chromosomal DNA into the break [14].

NHEJ is also the mechanism by which transgene integration occurs following either *Agrobacterium*-mediated or direct transformation of plant cells. The integration sites are generally random, but transcriptionally active sites seem to be preferred.

When we consider the evolution of gene targeting research, HR pathways were initially considered the only route with potential to achieve this because of high levels of fidelity observed in HR during meiosis. The levels of homology involved in meiotic recombination are large and would make this approach unworkable for routine plant genetic engineering. The extent of homology required is extensive, and may elongate the transgenes required in plant transformation to impractical levels. Induction of double-stranded breaks on the DNA by exposure to X-rays or by transposon activity was shown to increase HR [15, 16]. Site-specific recombination systems therefore became a potential route to achieving gene targeting by HR, since they can introduce double-stranded breaks in DNA, and repair these in via an HR mechanism that utilizes shorter homologies.

The objective of many plant transformation research groups is to study genomics and generate improved crops. While transgenic plants produced for genomics study have little regulatory requirements since they are for contained use, transgenic plants for general release have to comply with governmental regulations and must also meet consumer acceptance. Gene targeting will make it easier for genetically modified plants to meet these requirements. A strategy for gene targeting that has been explored extensively by researchers is that of site-specific recombination.

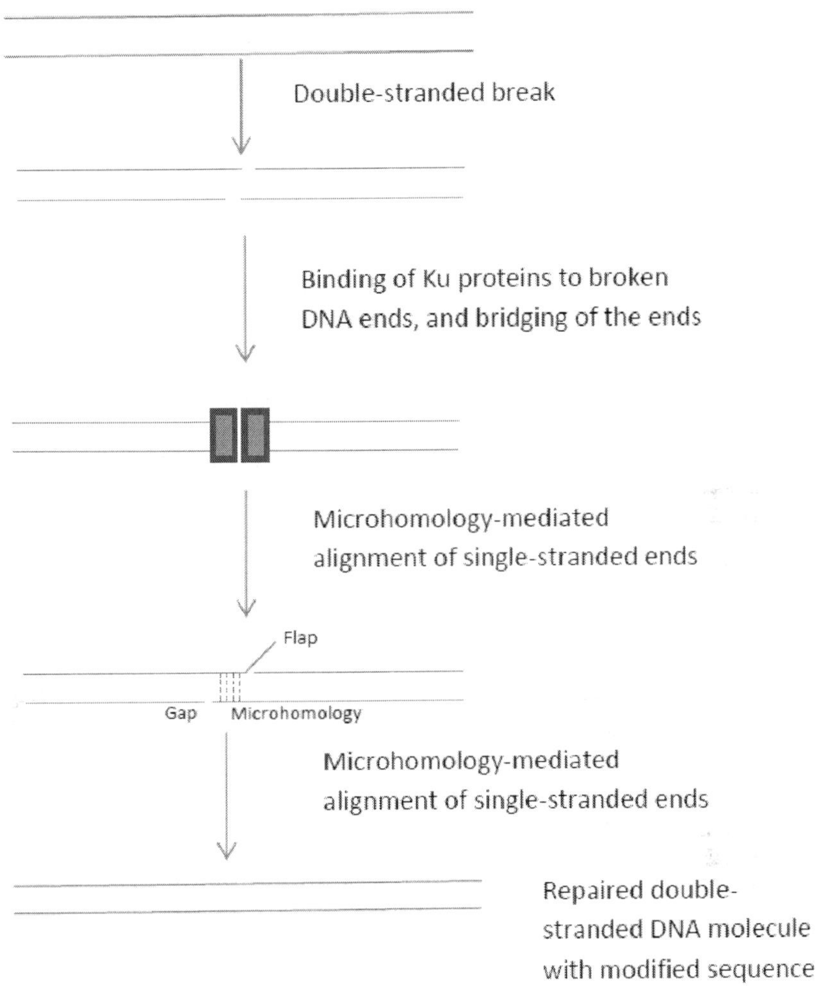

Figure 2. Model for repair of double-stranded breaks by non-homologous end-joining. Adapted from [8].

Site-specific recombination systems consist of a recombinase and donor sites. The recombinase is a protein that mediates a recombination reaction between a target site characterized by particular target sequence for that protein, and the donor site, also with a characteristic nucleotide sequence. In general, the results of the ensuing recombination reaction are excision, integration or inversion.

The site-specific recombination systems that can be utilized for gene targeting include the tyrosine family recombination system, the serine family recombination system and the newly developed hybrid system consisting of zinc finger DNA sequence recognition motifs in combination with a rare-cutting restriction endonuclease. Each of these systems will now be considered in turn, and the potential to contribute to gene targeting discussed.

3. Tyrosine family recombination systems

The tyrosine family recombination systems include the Cre/*lox P*, FLP/*FRP*, λ integrase and variations thereof [17]. Also known as integrases, they use the hydroxyl group of the catalytic tyrosine for a nucleophilic attack on the phosphodiester bond of the target DNA, and function through a Holliday junction intermediate [17, 18]. Their function depends on the cofactors supplied.

Cre and FLP recombinases are the most popular members of the integrase family because they are simple and unrestrictive, requiring no auxillary factors other than their recombinase monomers and their cognate targets. Cre recombinase recombines 34 bp *lox P* sites in the absence of accessory proteins or auxillary DNA sequences [17, 19]. The FLP target site has been trimmed from the original 599 bp in the 2 μm yeast plasmid to 30 bp in *FRT* sites [17]. The wild-type *FRT* and *lox P* sites are unchanged by the recombination reaction, making the reaction reversible; and there are many different possible recombination intermediates in each case. This has however made it difficult to utilize the tyrosine family recombinase systems more extensively in vector construction, while the irreversible λ integrase is more popularly used in vectors [17].

4. Serine family recombinase systems

The Serine family recombinase systems such as ϕC31, Hin and Gin [17] are also known as the resolvases or invertases. They have a conserved serine residue that is used to create the covalent link between the recombinase and the DNA target site [18]. Serine family recombinases initiate strand-exchange by making double-stranded breaks at two sites in the DNA molecules. Each site of the double-stranded break is associated with a dimer of the recombinase, and the two dimers will come together bringing the two broken ends together and forming an active tetramer in a process that is elaborately controlled [20].

The general scheme for using site-specific recombination systems in gene targeting involves, first, the genetic engineering of the recombination target sites into the particular genomic location of the plant to be transformed. This can be achieved by standard transformation procedures followed by screening to identify transformation events in 'acceptable' locations. Transposon tagging has also been used with the recombination target sites incorporated within the transposon.

The second requirement is that the incoming transgene should have unique DNA sequences that constitute the donor sites. Finally, there should be a mechanism for expression or introduction of the recombinase, to mediate the recombination reaction between donor and target sites. In this scheme, a second transformation experiment targets the genes into which the recombinase target sequences were integrated by the first transformation experiment.

Both *Agrobacterium*-mediated and direct gene transfer (bombardment, electroporation and PEG-mediated transformation) have been used for the initial transformation to introduce target and donor sites. The recombinase may be expressed constitutively, transiently or may be induced. Recombinase expression as well as stability of the transgene may vary [21].

These approaches were based on the need to improve homologous recombination at the target site. In these approaches, homology is limited to target and donor site compatibility for the particular recombinase being considered. With elegant engineering, site-specific recombination systems can be used to remove marker genes from transgenic plants before their commercialization. But the process is far from routine. Also, the footprint that remains on the chromosome is associated with genetic instability. The search for a better system continues, and that is why zinc finger nucleases are being considered.

5. Zinc finger nuclease and gene targeting

Zinc finger nucleases (ZFN) are artificial restriction endonucleases composed of a fusion between an artificial Cys_2His_2 zinc finger protein DNA binding domain and the cleavage domain of the *Fok* I endonuclease. The sequence-specific DNA binding domain of zinc fingers could be engineered to recognize a variety of specific DNA nucleotide sequences of the researcher's choice [22]. While the *Fok* I endonuclease activity is non-specific, the enzyme only functions when it forms a dimer, whose assembly will be guided by proper alignment of the two zinc finger monomers at the target site. Assembly of the ZFN therefore enables site-specific cleavage [3, 23]. The two zinc finger monomers are usually designed to flank a 5 – 6 bp long target sequence, within which *Fok* I cleavage will occur. The zinc finger domain itself is composed of 3 – 4 individual fingers, each of which recognizes 3 bp sequences [1]. Overall, a unique sequence of about 24 bp is specifically recognized, and this is large enough to be unique in most genomes.

The most common forms of the ZFN recognition sites are $(NNY)_3N_6(RNN)_3$, of which $(NNC)_3N_6(GNN)_3$ has been extensively studied [23, 24]. The double-stranded breaks will significantly increase integration of DNA into the target site by HR by up to 100 times in plants [1]. But even then, double-stranded breaks induced by restrictions endonucleases or transposons have been shown to be predominantly repaired by NHEJ, often accompanied by some level of mutagenesis [1, 7]. A high proportion of the double-stranded breaks will therefore be repaired by NHEJ, since it is the predominant repair mechanism in plants.

Once the double-stranded break is made, early approaches were to try and increase the chances of their being repaired by HR, over the more predominant NHEJ. The approach has not been very successful. Research efforts should rather focus on ensuring that the repair by NHEJ does not mutate the nucleotide sequence of the target gene in an undesirable manner.

There have been attempts to increase the chances of HR after inducing double-stranded breaks with ZFN. For example, use of ZFN in combination with recombinases and chromatin remodeling proteins, this system increases both targeting precision and transformation efficiencies by HR. Further development of the system should optimise removal or exclusion of marker and reporter genes as well as vector backbone sequences.

Gene targeting using ZFN was first demonstrated for the yellow locus in *Drosophila* [25]. The approach has since become standard for many animal species having been demonstrated even in humans [26]. In higher plants, the technology lagged behind, but has shown a lot of promise with for instance the use of a novel TRV-based vector to achieve non-transgenic genome modification in plant cells [27]. ZFN may also be used for gene deletion [28], and removal of marker genes. Genetically engineered plants released into the environment should not have unnecessary transgene sequences.

Failure to increase HR in plants does not mean that all is lost. In fact, maybe plant transformation efforts may well benefit from NHEJ, which is the predominant mechanism of recombination in plants cells anyway. If one considers for instance a scenario where one needs to disrupt an endogenous gene whose phenotype is easily assayed for. Transient expression of a ZFN that targets the gene should introduce double-stranded breaks in the gene, and most of the breaks will be repaired by NHEJ. Errors introduced during the repair process should inactivate the gene. The precisely engineered plant you get is non-transgenic because there are no foreign sequences integrated into its genome, but the genome would have been elegantly edited! Sequencing of the edited gene should be used to confirm and characterize the mutation. Marton and coworkers reported on a successful experiment with this approach using a disarmed *Tobacco rattle virus* (TRV) vector to deliver the ZFN to cells of intact tobacco and petunia plants [27]. The mutations that were induced were stable and heritable.

There are many other possibilities for gene targeting in plants. For instance, the efficacy of oligonucleotide-directed plant gene targeting has been demonstrated, again with the possibility of the plants being considered non-transgenic [29].

6. Prospects for further development

The efficacy of gene targeting in plants has now been demonstrated, and genetically engineered plants using this technology are being developed. These plants are expected to be low copy number, reflecting on the target gene, and in genomic locations that correspond to the natural locations of the targets. Gene targeting approaches utilize the vast amount of genomic data that is now readily available in databases and can be correlated with the stability of the modification introduced at particular genomic sites. This would be ideal to enhance agricultural attributes of crop, for instance by increasing the expression of a desired product or shutting down a competing or undesirable pathway. With the levels of precision and true engineering that comes with gene targeting, the dependency on reporter genes and even marker genes is reduced. New and elegant ways of delivering DNA to plant cells, such as oligonucleotides and minimal cassettes will enable plant transformation without the use of plant transformation vectors whose backbones are notorious for integrating into the plant genome. Marker-free, vector backbone-free precisely engineered agricultural crops are what farmers, consumers and the environment need.

7. Conclusions

Gene targeting technology in plants has come a long way, and several alternative approaches to gene targeting have been evaluated. It is now possible and desirable for new plant transformation experiments to give some consideration as to which region of the genome they would want to target, and also give special consideration to reporter genes, marker genes and vector backbone sequences that might be associated with the experiment. It is hoped that the dream of reporter-free, marker-free, vector backbone-free truly and precisely engineered plants will soon be a reality.

Author details

Richard Mundembe*

Address all correspondence to: rmunde01@yahoo.com mundember@cput.ac.za

The Biotechnology Programme, Department of Agricultural and Food Sciences, Cape Peninsula University of Technology, Cape Town, South Africa

References

[1] Weinthal, D, Tovkach, A, Zeevi, V, & Tzfira, T. (2010). Genome editing in plant cells by zinc finger nucleases. Trends in Plant Science , 15, 308-321.

[2] Iida, S, & Terada, R. (2005). Modification of endogenous natural genes by gene targeting in rice and other higher plants. Plant Molecular Biology , 59, 205-219.

[3] Durai, S, Mani, M, Kandavelou, K, Wu, J, Porteus, M. H, & Chandrasegaran, S. (2005). Zinc finger nucleases: custom-designed molecular scissors for genome engineering of plant and mammalian cells. Nucleic Acids Research , 33(18), 5978-5990.

[4] Terada, R, Johzuka-hisatomi, Y, Saitoh, M, Asao, H, & Iida, S. (2007). Gene targeting by homologous recombination as a biotechnological tool for rice functional genomics. Plant Physiology , 144, 846-856.

[5] Evans, M. J, Smithies, O, & Capecchi, M. R. (2001). Mouse gene targeting. Nat Med , 7, 1081-1090.

[6] Puchta, H. (1999). DSB-induced recombination between ectopic homologous sequences in somatic plant cells. Genetics , 152, 1173-1181.

[7] Putcha, H. (2005). The repair of double-strand breaks in plants: mechanisms and consequences for genome evolution. Journal of Experimental Botany , 56(409), 1-14.

[8] Hefferin, M. L, & Tomkinson, A. E. (2005). Mechanism of DNA double-strand break repair by non-homologous end joining. DNA Repair , 4, 639-648.

[9] Keeney, S. (2001). Mechanism and control of meiotic recombination initiation. Current Topics in Developmental Biology , 52, 1-53.

[10] Shaley, G, & Levy, A. A. (1997). The maize transposable genetic element Ac induces recombination between donor site and a homologous ectopic sequence. Genetics , 146, 1143-1151.

[11] Fishman-lobell, J, Rudin, N, & Haber, J. E. (1992). Two alternative pathways of double-strand break repair that are kinetically separable and independently modulated. Molecular and Cellular Biology , 12, 1292-1303.

[12] Orel, N, Kirik, A, & Puchta, H. (2003). Different pathways of homologous recombination are used for repair of double-stranded breaks within tandemly arranged sequences in the plant genome. The Plant Journal , 35, 604-712.

[13] Nicolas, A. L, Munz, P. L, & Young, C. S. (1995). A modified single-strand annealing model best explains the joining of DNA double-strand breaks in mammalian cells and cell extracts. Nucleic Acids Research , 23, 1036-1043.

[14] Salomon, S, & Puchta, H. (1998). Capture of genomic and T-DNA sequences during double-strand break repair in somatic plant cells. EMBO Journal , 17, 6086-6095.

[15] Tovar, J, & Lichtenstein, C. (1992). Somatic and meiotic chromosomal recombination between inverted duplications in transgenic tobacco plants. The Plant Cell , 4, 319-332.

[16] Xiao, Y. L, & Peterson, T. (2000). Interchromosomal homologous recombination in Arabidopsis induced by a maize transposon. Molecular and General Genetics , 263, 22-29.

[17] Lyznik, L. A, Gordon-kamm, W, Gao, H, & Scelonge, C. (2007). Application of site-specific recombination systems for targeted modification of plant genomes. Transgenic Plant Research , 1(1), 1-9.

[18] Grindley, N. D, Whiteson, K. L, & Rice, P. A. (2006). Mechanisms of site-specific recombination. Annual Reviews in Biochemistry , 75, 567-605.

[19] Gopaul, D. N, Guo, F, & Van Duyne, G. D. (1998). Structure of the Holliday junction intermediate in Cre-loxP site-specific recombination. The EMBO Journal , 17(14), 4175-4187.

[20] Marshall, S. W, Boocock, M. R, Olorunniji, F. J, & Rowland, S. J. Intermediates in serine recombinase-mediated site-specific recombination. Biochemical Society Transections , 39(2), 617-622.

[21] Kumar, S, Franco, M, & Allen, G. C. (2006). Gene targeting: Development of novel systems for genome engineering in plants. Floriculture, Ornamental and Plant Biotechnology Volume IV, Chapter 8, Global Science Books., 84-98.

[22] Fauser, F, Roth, N, Pacher, M, & Ilg, G. Sanchez-Fernandez, Biesgen, C. and Puchta, H. ((2012). In plant gene targeting. PNAS , 109(19), 7535-7540.

[23] Carroll, D, Morton, J. J, Beumer, K. J, & Segal, D. J. (2006). Design, construction and in vitro testing of zinc finger nucleases. Nature Protocols , 1, 1329-1341.

[24] Liu, Q, Xia, Z. Q, Zhong, X, & Case, C. C. (2002). Validated zinc finger protein designs for all 16 GNN DNA triplet targets. Journal of Biological Chemistry , 277, 3850-3856.

[25] Bibikova, M, Carroll, D, Segal, D. J, Trautman, J. K, Smith, J, Kim, Y-G, & Chandrasegaran, S. (2001). Stimulation of homologous recombination through targeted cleavage by chimeric nucleases. Molecular and Cell Biology , 21, 289-297.

[26] Porteus, M. H, & Carroll, D. (2005). Gene targeting using zinc finger nucleases. Nature Biotechnology , 23, 967-973.

[27] Marton, I, Zuker, A, Shklarman, E, Zeevi, V, Tovkach, A, Roffe, S, Ovadis, M, Tzfira, T, & Vainstein, A. (2010). Non-transgenic genome modification in plant cells. Plant Physiology , 154, 1079-1087.

[28] Petolino, J. F, Worden, A, Curlee, K, Connell, J, Moynahan, T. L. S, Larsen, C, & Russell, S. (2010). Plant Molecular Biology , 73, 617-628.

[29] Oh, T. J, & May, G. D. (2001). Oligonucleotide-directed gene targeting. Current Opinion in Biotechnology12: 169 172.

Section 3

Genetic Engineering and Cloning in Animals

Chapter 4

Genetic Engineering and Cloning: Focus on Animal Biotechnology

Mariana Ianello Giassetti*,
Fernanda Sevciuc Maria*,
Mayra Elena Ortiz D'Ávila Assumpção and
José Antônio Visintin

Additional information is available at the end of the chapter

http://dx.doi.org/10.5772/56071

1. Introduction

1.1. What is genetic engineering?

Over the last 35 years the term genetic engineering has been commonly used not only in science but also in others parts of society. Nowadays this name is often associated by the media forensic techniques to solve crimes, paternity, medical diagnosis and, gene mapping and sequencing. The popularization of genetic engineering is consequence of its wide use in laboratories around the world and, developing of modern and efficient techniques. The genetic engineering, often used with trivia, involves sophisticated techniques of gene manipulation, cloning and modification. Many authors consider this term as synonymous as genetic modification, where a synthetic gene or foreign DNA is inserted into an organism of interest. Organism that receives this recombinant DNA is considered as genetically modified (GMO). Its production are summarized in simplified form in five steps: 1) Isolation of interested gene, 2) Construction, gene of interested is joined with promoters (location and control the level of expression), terminator (indicates end of the gene) and expression marker (identify the gene expression), 3) transformation (when the recombinant DNA is inserted into the host organism), 4) Selection (selection of those organisms that express the markers), 5) Insertion verification of recombinant DNA and its expression [1].

1.2. How to apply genetic engineering in our everyday

One of the main firstlings of genetic engineering is that genetic information is organized in the form of genes formed by DNA, which across some biotechnologies can be manipulated to be

INTECH
open science | open minds

© 2013 Giassetti* et al.; licensee InTech. This is a paper distributed under the terms of the Creative Commons Attribution License (http://creativecommons.org/licenses/by/3.0), which permits unrestricted use, distribution, and reproduction in any medium, provided the original work is properly cited.

applied in various fields of science. Currently, genetic engineering is widely used at various branches of medicine to produce vaccines, monoclonal antibodies, animals that can be used as models for diseases or to be used as organ donors (such as pigs). Another function of genetic engineering is gene therapy which aims to restore correct gene expression in cells that have a defective form. In the industry, genetic engineering has been extensively used for the production bioreactor able to express proteins and enzymes with high functional activity. Already in agriculture, genetic engineering is being very controversial because it tends to produce genetically modified foods resistant to pests, diseases and herbicides.

1.3. Concept is already old

However, all the knowledge obtained in the present day was only possible by discoveries of Gregor Mendel, considered the father of genetics. The results obtained in 1865 by the Austrian monk generated genetics studies related to heritability and variation. The term formerly called "element" by Mendel was later termed "genes" by Wilhelm Johanssen in 1909. Sutton and Boveri (1902) have proposed that these genes were grouped in the form of chromosomes, which in turn constitute the genetic material of eukaryotes. In 1953, James Watson and Francis Crick unraveled the structure of DNA as double helix, creating a period of intense scientific activity that culminated in 1966 with the establishment of the complete genetic code.

Major new discoveries were made in 1967 when DNA ligase was isolated that has the ability to join DNA fragments. The first restriction endonuclease enzyme was isolated in 1970 and it functions as a scissors cutting a specific DNA sequence. These discoveries allowed the development of the first recombinant DNA molecule, which was first described in 1972. In 1973 restriction enzymes (scissors) and DNA ligase (adhesive) were used to join a DNA fragment in plasmid pSC101, which is a circular extrachromosomal bacterial DNA. Thus, E. coli was transformed with the recombinant plasmid and it was replicated, generating multiple copies of the same recombinant DNA. The experiments conducted in 1972 and 1973 were crucial to the establishment of new genetics and genetic engineering.

1.4. Genome: Structure, organization and function

Genome is considered long chains of nucleic acid that contains the information necessary to form an organism [2], consisting of small subunits called nucleic bases that are inheritable. Thus, the genome contains a complete set of features that are inheritable. The genome can be divided functionally into sets of base sequences, called genes. Each gene is responsible for coding a protein, and alternative forms called alleles. A linear chain gene is named chromosome and each gene assumes a specific place, locus. Therefore, the modern view of genetics genome is a complete set of chromosomes for each individual. According to the central dogma (Figure 1), each gene sequence encodes another sequence of nitrogenous bases of single stranded RNA. The RNA sequence, complementary to a genomic DNA, will encode amino acids that form the protein. As previously mentioned, each gene relates with expression of one protein and for that each codon (the sequence of 3 nitrogenous bases of DNA) represent only one amino acid, but each amino acid can be represented by more than one codon.

http://dx.doi.org/10.5772/56071

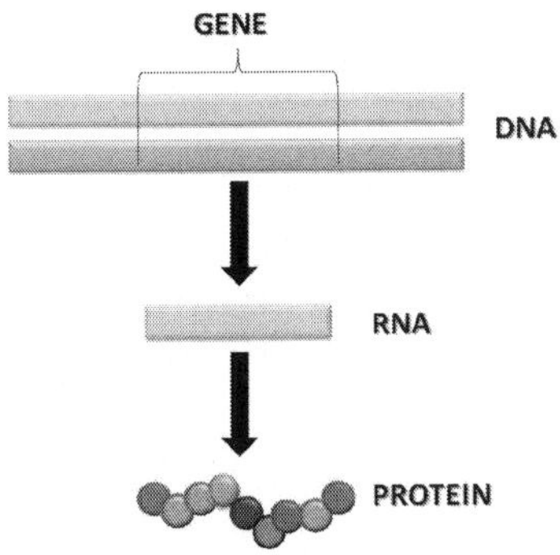

Figure 1. Central Dogma, gene codes a RNA sequence that is complementary of DNA and it encodes a protein.

1.5. DNA and RNA structure

The DNA is considered as genetic material of bacteria, viruses and eukaryotic cells having a basic structure the nucleotide, which is formed by a nitrogenous base (purine ring or pyrimidine), sugar and phosphate. In 1953, Watson and Crick proposed that DNA is a double polynucleotide chain organized as a double helix. In this model, the double helix was linked by hydrogen bounding between nitrogenous bases. The base is linked to the 1-position by a pentose glycosidic bond from N7 of pyrimidines or N9 of purine. The nuclear acid is named by the type of sugar. DNA has 2`-deoxyribose, whereas RNA has ribose. The sugar in RNA has an OH group in a 2` position of pentose ring. A nucleic acid is a long chain of nucleotides and the sugar can be linked in 3′or 5′ position to the phosphate group and the backbone of chain consist in a repeated sequence of sugar (pentose) and phosphate residues. One pentose ring is connected at 5`position to a forward pentose that is linked by the 3` position via phosphate residues; in this way, the sugar-phosphate backbone is 5′-3` phosphodiester linkages (Figure 2).

Nucleic acid contains 4 types of base, 2 purines (adenine (**A**) and guanine (**G**), which are present in DNA and RNA) and two pyrimidines (cytosine (**C**) and thymine (**T**) for DNA and for RNA uracil (**U**) instead of thymine). Therefore, DNA contains **A**, **G**, **C**, **T** and RNA contains **A**, **G**, **C** and **U**. Other important discover were that the **G** bounded specifically with **C**, and **T/U** with **A**; these named base pairing (complementary), and that the chains had apposite directions (antiparallel).

2. Genetic engineering: Timeline

The chronological order of main events of genetic engineering and cloning are described above.

1866 - Gregor Mendel proposed the law of independent, of segregation and basic principles of heredity; principles that created a new science called genetic.

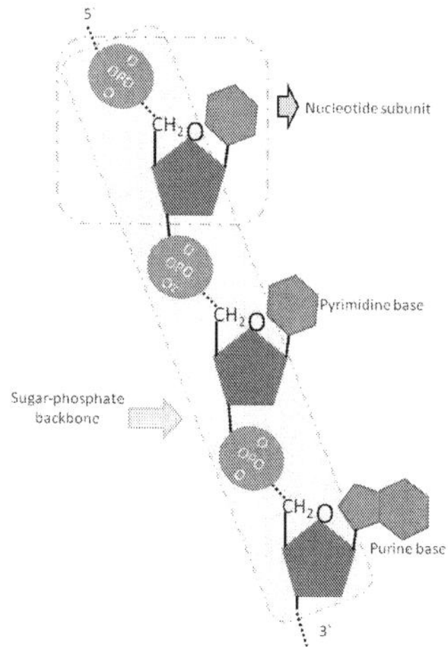

Figure 2. Polynucleotide chain, 5´-3´sugar phosphate linkages (backbone) and structure of nucleotide subunit - Adapted from Lewin, B (2004)[2]

1900 - Mendel´s principles were rediscovered by Hugo de Vries, Carl Correns and Eric von Tschermak

1908 - Chromosome Theory of Heredity was proposed by Thomas Hunt Morgan

1944 - Was established that DNA contains the heredity material.

1946 - First electronic digital computer was created

1952 - The first cloned animal (Northern Leopard Frog)

1953 - Watson and Crick described DNA structure and proposed the double helix model.

1955 - Protein sequencing method was established by Frederick Sanger and insulin was sequenced

1965 - Atlas of protein sequences was created

1966 - Genetic code was cracked

1970 - Algorithm for DNA sequence was created

1972 - Establishment of DNA recombinant technology by Stanley Cohen, Herbert Boyer and Paul Berg

1973 - The first recombinant DNA organism was created

1976 - The first genetic engineering company is founded.

1977 - DNA sequencing method was established

1980 - Was done the first molecular mapping of a human chromosome

1982 - GeneBank started to be public

1983 - Mullis developed PCR (Polymerase chain reaction)

1984 - "Genetic Fingerprinting" techniques was developed and human genome sequencing started

1986 - National Center for Biotechnology was developed in USA and automatic machine for DNA sequencing was created

1990 - Dolly, the first cloned animal, was born and blast program was created

1995 - First complete bacterial genome was sequenced

1997 - E. coli complete genome sequence was published

1999 - Complete sequence of human chromosome 22 was published

2000 - Drosophila genome was sequenced and first holy genome from plant was published

2002 - Mouse genome sequence was published

2003 - Human genome sequence was published

2004 - Chimpanzee genome sequence was published

3. Cutting and pasting the DNA

3.1. Discovering restriction endonuclease and a Nobel Price in 1978.

Molecular biology and genetic were innovated in middle of 70th decade the discover of restriction endonuclease by W Arber, D Nathans e H Smith that wan the Nobel price in 1978. When phage λ attacks an E. coli strain B a specific restriction endonuclease (EcoB) cuts just the DNA from phage λ and infections is blocked. E.coli methylates its own DNA by action of DNA methylase to protect this DNA from itself enzyme. Restriction endonuclease recognizes short sequences of duplex DNA as cleavage target and the enzyme cuts this point of DNA every time this target sequence occurs. When the DNA molecule is cleaved by restriction endonuclease DNA fragments are produced. Analyzing restriction fragments is possible to generate a map of the original DNA molecule (restriction map, a linear sequence of DNA separated in defined fragment size) [1, 2]

3.2. Types of restriction endonuclease enzyme: Nature, structure, application, recognition site of action and nomenclature

Restriction endonuclease are classified in types I, II and III by sequence specificity, nature of restriction and structural differences (table 1). Types I and III have a restrict use in molecular biology and genetic engineering but the type II is largest used because it cleaves the DNA a specific recognition sequence, separate methylation, no additional energy requirement is necessar, high precision and do not match actions. Type II restriction endonuclease are classified by the size of recognition sequence such as tetracutter, hexacutter or octacutter (4, 6 and 8 base paired respectively) [3].; and generally that sequences are palindromic (nitrogenous bases sequence read the same backwards and forwards). Restriction enzymes also could be classified as neoschizomers (recognize the same sequence) and isoschizomers (recognize and cleave in the same location).

	Type I	Type II	Type III
Enzyme structure	Complex of three subunits with independent recognition endonuclease and methylase function	Separate monomeric enzymes for endonuclease and methylase action, both recognize the same target sequence	Separated dimeric enzymes for endonuclease and methylase with one common subunit
Requirement for activation	ATP and Mg2+ S-Adenosyl methionine	Mg^{2+}	ATP and Mg2+ S-Adenosyl methionine Enhance activity
Recognition site	Double-stranded DNA	Generally palindromic sequence of Double-stranded DNA	Single-stranded DNA
Nature of restriction	Cleaves the DNA at a random sequence at one Kb away of recognition site	Cleaves the DNA at a specific sequence near or at the recognition site	Cleaves the DNA about 25pb downstream of the recognition site at a random sequence

Table 1. Properties of restriction endonucleases – Adapted from Satya, P[3] (2007)

The nomenclature of restriction endonuclease is derivate from the species that it was isolated (Ex. ECORI, from *Escherichia coli Ry13*); First two letters from enzyme name identify the species and the third identify the different strains from the same organism (Table 2). The number classifies the different enzymes from the same organism and strains in chronological order of discover (Ex. Hind III, is the third RE isolated from *Haemophilus influenza*). Restriction endonuclease cut the DNA in two different ways: blunt end (two DNA strands are cleaved at the same position) or sticky end (the enzyme cut each DNA strand at different position, generally two until four nucleotides apart). So in the sticky, DNA fragments have short single-stranded overhangs at each end. *) [1-4]*

3.3. Linking of DNA fragments: DNA ligase

Restriction endonuclease type II cuts the double-stranded DNA in specific target sequence but this enzyme do not joined back again the DNA fragments, this is essential to create a new hybrid DNA. Joining two DNA fragments by 5`→3` phosphodiester bond is an energy dependent process (ATP or NAD, depending the kind of enzyme that is being used). DNA ligase is a specific enzyme that is responsible to join DNA fragments spending and two blunt

Enzyme	Recognition sequence	Type of ends	End sequences
*Alu*I	5′-AGCT-3′	Blunt	5′-AG ↕ CT-3′
	3′-TCGA-5′		3′-TC ↕ GA-5′
*Sau*3AI	5′-GATC-3′	Sticky, 5′ overhang	5′- ↕ GATC-3′
	3′-CTAG-5′		3′-CTAG ↕ -5′
*Hinf*I	5′-GANTC-3′	Sticky, 5′ overhang	5′-**G** ↕ ANTC-3′
	3′-CTNAG-5′		3′-CTNA ↕ **G**-5′
*Bam*HI	5′-GGATCC-3′	Sticky, 5′ overhang	5′-**G** ↕ GATCC-3′
	3′-CCTAGG-5′		3′-CCTAG ↕ **G**-5′
*Bsr*BI	5′-CCGCTC-3′	Blunt	5′- ↕ NNNCCGCTC-3′
	3′-GGCGAG-5′		3′- ↕ NNNGGCGAG-5′
*Eco*RI	5′-GAATTC-3′	Sticky, 5′ overhang	5′-**G** ↕ AATTC-3′
	3′-CTTAAG-5′		3′-CTTAA ↕ **G**-5′
*Pst***I**	5′-CTGCAG-3′	Sticky, 3′ overhang	5′-CTGCA ↕ **G**-3′
	3′-GACGTC-5′		3′-**G** ↕ ACGTC-5′
*Not***I**	5′-GCGGCCGC-3′	Sticky, 5′ overhang	5′-GC ↕ GGCCGC-3′
	3′-CGCCGGCG-5′		3′-CGCCGG ↕ CG-5′
*ggl***I**	5′-GCCNNNNNGGC-3′	Sticky, 3′ overhang	5′-GCCNNNN ↕ NGGC-3′
	3′-CGGNNNNNCCG-5′		3′-CGGN ↕ NNNNCCG-5′

*N = any nucleotide.

*Note that most, but not all, recognition sequences have inverted symmetry: when read in the 5′→3′ direction, the sequence is the same in both strands.

Table 2. Same restriction endonuclease used in genetic engineering - Adapted from Brown, TA (2002)[4]

ends can be joined easily spending two ATPs molecules and this blunt end is very popular in genetic engineering. However, the efficiency of this process is very low because the DNA ligase just joins adjacent DNA fragments and it cannot bring the DNA end nearby. Action of enzyme to catalyze the reaction is a random process that depends of vicinity of DNA fragments in solution. Joining DNA fragments with blunt ends is generally used to short oligonucleotides because concentration of free ends and enzyme are high, increasing the efficiency of process. Presence of sticky ends increase process efficiency because complementary ends come together by a random diffusion event in the solution and transient base pair might form between the two complementary strand. This ligation is not very stable but may persist for enough time to join DNA fragments by DNA ligase catalysis and synthesis of phosphodiester bonds [4].

The greater efficiency of sticky-end ligation stimulated the creation of new methods, such linkers or adaptors. They are short double-strand molecules that cover the blunt-end and insert a recognition sequence for a restriction endonuclease to create a sticky-end. The linkers need to be digest by a restriction endonuclease to have a stick-end but the adaptor is a final sequence, digestion is not necessary and fragments can be direct joined by DNA ligase.

4. DNA cloning

In modern molecular biology the ability to manipulate DNA molecules by restriction endonuclease and DNA ligase is named by DNA cloning and, a recombinant DNA can be constructed. However, a single copy of recombinant DNA is not enough. Replication machinery of one organism generally is used to increase the number of copies. The DNA is inserted in the organism for a propagation or transfer. Generally, the vector has autonomic replication system that is independent of the cell cycle, increasing the number of copies. Majority systems of DNA cloning use bacterial as a host and common plasmid vector is classified in low copy number (<10) or high copy number (>20). To select recombinant cell some parameters need to be present: have restriction sites in which de exogenous DNA is inserted (just one site for each restriction endonuclease) and vector needs to have a marker gene multicloning sites (one site for several restriction endonuclease) makes the vector more useful [3, 4].

5. Isolation, sequencing and synthesis DNA

The transgenic animal technology involves in first place, the isolation or artificially synthesis of a gene, which will be molecular manipulated and used for transformation leading to the transgenic production. The need of knowledge involving this target gene can be overcome by its sequencing, conducting to the understanding of its structure. The main of this subject is briefly described the mechanisms involved in isolation, sequencing and synthesis of a gene.

5.1. Isolation of genes

The first gene isolation was reported in 1969. Two specialized transducing phages, bacteriophages λ and Φ80, which carry the *lac* operon of *Escherichia coli* was inserted in reverse orientation into their DNA, being used as a source of complementary sequences to prepare pure *lac* operon duplex [5]. This method besides being resourceful work did not have general applicability.

Now a day, several methods are in progress for isolation of a gene. A most traditional method used largely in research is the construction of a genomic or complementary DNA (cDNA) library. A genomic library represents the total DNA of a cell including the coding and non-coding sequences cloned on a vector and a cDNA library is a combination of cloned fragments from the mRNA inserts into a collection of host cells, creating in both cases, a portion of organism transcriptome.

To produce a genomic library after the extraction of genomic DNA, these molecules are digested into fragments of reasonable size by restrictions endonucleases and then inserted into a cloning vector generating a population of chimeric vector molecules.

On the other hand, to create a cDNA library is necessary first, to produce a cDNA, which can be obtained from a mature mRNA isolate from a tissue or cells actively synthetizing proteins. The extraction of mRNA is easy due to poly-A tail present in eukaryotic mRNAs. Than the extracted mRNA is used for copying it into cDNA using the reverse transcriptase enzyme, method that create a single strand cDNA, which is converted in a double strand cDNA with DNA polymerase, coiling and nucleases. This cDNA are cloned into a bacterial plasmid, which is transformed into bacterial competent cells, amplified and selected.

Once a genomic or a cDNA library is available, they can be used for the identification and isolation of a gene sequence.

There are many commercial kits to create a genomic or cDNA library. Normally, the genomic library is created with lambda or cosmid vectors while a cDNA library is produced with plasmid vectors (*more information see item 5*). These kits usually try to improve the classical laborious techniques, enabling rapid construct of the libraries and ensuring generating of full-length clones.

The isolation of a gene using a genomic or cDNA library can be done by colony hybridization. In this technique the fragments containing a gene or parts of it can be identified by the use of DNA probes, which can be tagged or labeled with a molecular marker of either radioactive or fluorescent molecules. The commonly used markers are phosphorus 32 and digoxigenin, a non-radioactive, antibody-based marker.

The DNA of bacteria carrying the chimeric vectors is fixed on the filter, which is hybridized with the labeled probe carrying a sequence related to the gene to be isolated. The colonies carrying moderate to high similarity to the desired sequence are detected by visualizing the hybridized probe via autoradiography or other imaging techniques. In this way, the original chimeric vectors carrying the target gene sequence can be recovered from original colonies and used for advance researches.

If the library available were in the form of phage particles, instead of colonies are plaques that can be hybridized in the same way described above for colonies. This method of identification and isolation of genes are called plaques hybridization.

5.2. Isolation of genes related to a protein

To identify a gene related to a protein the inverse pathway (from protein to DNA) should be simulated. For start is necessary to have the protein product in a pure form. To purify a protein several methods typically used are in a series of steps. Each step of protein purification usually results in some degree of product loss, so, an ideal strategy is one in which the highest level of purification is reached in the fewer steps. The properties of the protein product like size, charge and solubility; determines the selection of which steps to use. These steps can be precipitation and differential solubilization; ultracentrifugation or chromatographic methods.

Thus, having the protein product is possible to produce antibodies probes for this protein by immunizing animals. This production require reliance upon animals immune system to levy responses that result in biosynthesis of antibodies against the inject molecule. Antigens must be prepared and delivered in a form and manner that maximizes production of a specific immune response by the animal.

These antibodies probes can be used to precipitation of polysomes engaged in synthesizing the target protein leading to the achievement of the mRNA coded for this protein. This method combined with immunoadsorbent techniques brings the possibility of application at less abundant proteins expression [6]. Then the mRNA are isolated and purified from the polysome fraction, being after used for synthesizing cDNA for a cDNA library preparation, described above.

Thereby, to identify the specific cDNA clone for the target protein immunological and electrophoretic analysis methods are used, screening a complete or partial genomic library [10].

5.3. DNA Sequencing

The basic concept of DNA sequencing is the mechanism involved in determining the order of nucleotides bases (adenine, guanine, cytosine and thymine) in a strand of DNA. F. Sanger and coworkers reported the first DNA sequencing, which was genome of DNA ΦX174 virus. Thus, at that moment, two methods of DNA sequencing were developed: one proposed by A. Maxam and W. Gilbert, known as chemical method of DNA sequencing, and the other developed by F. Sanger, S. Nicklen and A. R. Coulson known as chain termination method.

The chemical method of DNA sequencing consists in determines the nucleotide sequence of a terminally labeled DNA molecule by breaking it at adenosine, guanine, cytosine and thymine with chemical agents. Partial cleavage at each base produces a nested set of radioactive fragments extending from the labeled end to each of the positions of the base. The autoradiograph of a gel produced from four different chemical cleavages, shows a pattern of bands from which the sequences are read directly [7].

The chain termination method depends on DNA replication and termination of replication at specific sequences. For that, Sanger's technique is based on an enzymatic synthesis from a single-stranded DNA template with chain termination on DNA polymerase, using dideoxynucleotides (ddNTPs). The principle of this method relies on the dideoxynucleotide lacking a 3'OH group, which is required for extension of the sugar phosphate backbone. Thus, DNA polymerases cannot extend the template copy chain beyond the incorporated ddNTP [3, 8].

Both methods rely on four-lane high-resolution polyacrylamide gel electrophoresis to separate the labeled fragment and allow the base sequence to be read in a staggered ladder-like fashion. Sanger sequencing was technically easier and faster, becoming the main basis of DNA sequencing, being modified and automated to aid large scale sequence procedure [3, 8, 9].

5.3.1. Automatic sequencing

An automatic sequencing is an improvement of Sanger sequencing, through the use of different fluorescent dyes incorporated into DNA extension products primers or terminator. The use of different fluorophores in the four based (A, C, G and T) specific extension reactions means that all reactions can be loaded in a single lane. For each base one color are used, emitting a different wavelength when excited. Throughout electrophoresis, the fluorescence signs are detected and recorded [10, 11].

The classic electrophoresis methods used in automated sequencing are slab gel sequencing system or capillary sequence gel system, both described below.

5.3.2. *Slab gel sequencing systems*

The slab gel sequencing system consists of using ultrathin slab gels, about 75μm, and comprises running of at least 96 lanes per gel. By this instrument, fluorescent-labeled fragments were loaded to the top of vertical gel and electric filed was applied, as the negatively charged DNA fragments migrated through the gel they were sized and fractionated by the polyacrylamide gel. The fragments were automatically excited with a scanning argon laser and detected by a camera [12].

The loading of sequencing gels samples can be done manually or automatically. The automation consists in the use of a plexiglass block with wells in same distance from each other as the comb teeth cut in a porous membrane used as a comb for drawing samples by capillary action. The loading of samples automation achieve up to 480 samples per gel [9].

5.3.3. *Capillary sequence gel systems*

Alternatively, the capillary sequence gel system instead of continuous polyacrylamide gel slabs, DNA is sent through a set of 96 capillary tubes filled with polymerized gel [3, 9].

In this system fused silica capillaries of 50-100 μm in diameter and 30-80 cm in length, heat resistant, are filled with a separation matrix consisting of a gel and electrode buffer. Solution phase DNA molecule are injected into the capillary either by pressure or electrokinetic injection and separated inside the capillary according to their size under high voltage conditions. The molecules are detected using UV light absorption or laser induced fluorescent detection at the end of the capillary [3, 12].

5.3.4. *Direct sequencing by PCR*

PCR has relieved much of the experimental toil of molecular biology improving the procedure's sensibility and facilitating the rapid cloning and sequencing of large numbers of samples [13]. The amplification of target DNA by PCR followed by direct sequencing of amplified DNA has emerged as a powerful strategy for rapid molecular genetics analysis bypassing the time consuming cloning steps and generating accurate DNA sequence information from small quantities of precious biological samples [14].

The direct PCR sequencing involves two steps 1- generation of sequencing templates through PCR and 2- sequencing of PCR products using thermolabile or thermostable DNA polymerases [15].

Some enzymes as *Taq* polymerase are thermostable and can be used in automated sequencing reactions such as cycle sequencing. Others, such as Klenow polymerase and reverse transcriptase are thermal instable, being able to both direct sequencing by PCR products and cloned template, although cannot be used in cycle sequencing. Another enzyme, Sequenase, has also been used effectively in both radioactive and fluorescence cycle sequencing [8].

One sequencing strategy of form any PCR-amplified DNA template are the sequenase approach. First, the PCR-amplified DNA is denatured to single strands, annealing the sequencing primer to complementary sequence on one of the template strands. Then, the annealed primer is extended by DNA polymerase by 20-80 nucleotides, incorporating multiple radioactive labels into the newly synthesized DNA, under non-optimal reactions conditions, retaining the enzyme functionality low, for the synthesis of only short stretches DNA. After, the labeled DNA chains are extended and terminated by incorporation of ddNMPs [14].

On the other hand, cycle sequencing strategies can be used for PCR-amplified DNA. These methods generate high-intensity sequence ladders due to the advantage of automated cycling capability of thermal cyclers. First, the PCR-amplified DNA is denatured to single strands, and then it is annealed of a 32p-labeled sequencing primer. After, it is extended and chain-terminated by a thermostable DNA polymerase and denatured in the next sequencing cycle. This step releases the template strand for another round of priming reactions while accumulates chain-terminated products in each cycle. These steps are repeated 20-40 cycles to amplify the chain-terminated products in a linear fashion [14].

5.3.5. DNA sequencing by microarray

A DNA microarray technology brings the possibility of large scale sequence analyses by generating miniaturized arrays of densely packed oligonucleotide probes [9, 16].

The word microarray has been derived from the Greek word *mikro* (small) and the French word *arrayer* (arranged). This technology can be described as an ordered array of microscopic elements on a planar surface that allows the specific binding of genes or gene products [17, 18].

The DNA sequencing by microarray uses a set of oligonucleotide probes to examine for complementary sequences on a target strand of DNA. Briefly, after cleavage DNA segments are hybridized to the definitely arranged probes on a gene chip, the detection is made with a light driven. Then, to reconstruct the target DNA sequence, the hybridization pattern is used. To analyze the data and determinate the DNA sequence specific software are used [3, 16].

The array elements react specifically with labeled mixtures, producing signals that reveal the identity and concentration of each labeled species in solution. These attributes provide miniature biological assays that allow the exploration of any organism on a genomic scale [17].

The array technology has been widely used in functional genomics experiments designed to study the functions and interactions of genes within the context of the overall genome distinct plant and animal species. To sequence a DNA fragment by microarray a series of laboratory procedures are involved, from RNA extraction, reverse transcription and tagging fluorescent hybridization to the end, which invariably introduce different levels of additional variation data. On the other hand, experiments with microarrays are still considerably expensive and laborious and, as a consequence, are generally conducted with relatively small sample sizes. Thus, the conducting tests on microarrays require careful experimental design and statistical analysis of the data [19].

5.3.6. DNA sequencing by MALDI TOF mass spectrometry

The Matrix assisted laser desorption/ionization is very rapid and combined with time-of-flight (MALDI-TOF) became an efficient and less time consuming (range of several microseconds) in acquire DNA sequence information by sensitive discrimination of their molecular masses.

The technique consists in embedded the samples to be analyzed in a crystalline structure of small organic compounds (matrix) and deposited on a conductive sample support. Then, the samples are irradiated with an ultraviolet (UV) laser with a wavelength of 266 or 337nm. The energy of the laser causes structural decomposition of the irradiated crystal and generates a particle cloud from which ions are extracted by an electric field. Following acceleration through the electric field, the ions drift through a field-free path and finally reach the detector. The results come from the calculation of ion masses by measuring their TOF, which is longer for larger molecules than for smaller ones. Due to single-charged, nonfragmented ions are mostly produced, parent ion masses can be determined from the resulting spectrum without the need for complex data processing. The masses are accessible as numerical data for direct processing and subsequent analysis [20].

The development of MALDI-TOF for an efficient DNA analyses happens due to needed of high throughput, parallel processing, simplified handling and low-cost techniques. The method uses an initial PCR amplification, which, PCR is carried out with a DNA polymerase that accepts ribonucleoside triphosphates (NTP) substrates. One of the four deoxynucleotides is replaced by an NTP. Fragments are generated by simple alkali backbone cleavage at the ribo-bases of the PCR products, generating oligonucleotide fragments each terminating with the ribonucleotide of the cycled primer extension reaction. Analysis is carried out by MALDI-TOF mass spectrometry. Differences between the unknown sample and a reference sequence are determined by changes in the results pattern [21, 22].

Nowadays, with the advent of genome sequencing projects been accomplished, sequences of DNA can be obtained and compared through electronically databases, than physically from clone libraries (described above). The available databases include locus information, organism species, the whole gene sequence, the reference authors and the status of the sequencing. The most used resource is the GenBank [23] provided for the *National Center for Biotechnology Information* (NCBI).

5.4. Synthesis

The gene synthesis methods had their main development during 1980s and 1990s. DNA gene synthesis is the process of writing the DNA. As DNA carries the genetic information of an organism, it could be viewed like a kind of information resource, enabling its reading (sequencing, described above) and writing (synthesis).

The oligonucleotides synthesis can be done rapidly and in high yields with different kinds of methods. The gene synthesis, together with the knowledge of full genomes, molecular cloning, and protein expression profiles, improved the biotechnology field, making possible to explore the whole functionality of an entire complex organism.

5.4.1. Gene synthesis machine

The gene synthesis machine is fully automated instrument, which synthesizes predetermined polynucleotide sequence. The principle involved is based on a combination of organic chemistry and molecular biological techniques.

Automatic gene machines, synthesize specific DNA sequences by programming the apparatus for the desired sequence. Briefly, the chosen sequence is entered in a keyboard and a microprocessor automatically opens the valve of nucleotide, chemical and solvent reservoir, controlling the whole process [15].

Containers of the four nucleotides (A, T, C and G) and reservoirs for reagent and solvent supports are connected with the synthesizer column. This column is packed with small silica beads, which provides support for assembly of DNA molecules. The desired sequence is synthesized on the silica beads which are later removed chemically [23].

Commercial services for gene synthesis are available from numerous companies worldwide. This gene synthesis method provides the possibility of creates entire genes without the need of a DNA template.

5.4.2. Gene synthesis from mRNA

The reports of a ribonuclease-sensitive endogenous DNA polymerase activity in particles of RNA tumor viruses by H.M. Temin and D. Baltimore enable the synthesis of complementary DNA (cDNA) using mRNA as template [9, 15, 24].

This enzyme, known as reverse transcriptase, are largely used in biotechnology research, and combined with the polymerase chain reaction create a methodology for DNA synthesis and amplification of the product.

To use the mRNA as a template first is necessary purify this molecule of the cell, or tissue. This can be done using oligo-dT cellulose spin columns, oligo-dT/ magnetic beads and coated plates. The principle involved at isolation of mRNA relies on base pairing between the polyA residues at the 3′ end of most mRNA, and the oligo (dT) residues coupled to the surface of cellulose spin columns or, magnetic beads or, a pre-coated 96 oligo-dT plate.

Independently of efficiency the three kinds of mRNA isolation are available commercially, facilitating the lab work.

Since the mRNA is available, the cDNA can be produced. To produce the cDNA the reaction should be done using mRNA template and a mix of, primers, reverse transcriptase, solution of four dNTPs and buffers. Depending on the experiment, ligo (dT)12-18, random hexanucleotides, or gene-specific antisense oligonucleotides can be used as primers for synthesis of first-strand cDNA [25].

The correctly native gene synthesis by this method depend on the fidelity of copying mRNA and also on the stability of DNA thus synthetized. Moreover, since mRNA of a gene does not have the complete transcript of the gene *in vivo* (intronic regions are dismissed) the synthesized gene will be smaller than the gene in vivo, but contain just the coding sequences, what could be a great advantage for research [9].

5.4.3. Synthesis by PCR

The gene synthesis by PCR, as described first for W. P. C. Stemmer and coworkers were reported having four steps. First the olygos are synthetized, and then the gene is assembled, amplified and cloned. Since single-stranded ends of complementary DNA fragments are filled in during the gene assembly process, cycling with DNA polymerase results in the formation of increasingly larger DNA fragments until the full-length gene is obtained [26].

The classical method involves the use of oligonucleotides of 40nt long that overlap each other by 20nt. The oligonucleotides are designed to cover the complete sequence of both strands, and the full-length molecule is generated progressively in a single reaction by overlap extension PCR, followed by amplification in a separate tube by PCR with two outer primers [27].

Variations of the classical approach were done, such as ligation of phosphorylated overlapping nucleotides, modified form of ligase chain reaction combinations with asymmetrical PCR and thermodynamically balanced inside out.

Nevertheless, most of them are based on phosphorylation of oligos at the 5′ ends', annealing of overlapping ends, filling the gaps by enzymatic extension at 3′ ends and join nicks with DNA ligase. Then the full length double stranded DNA can be cloned on a plasmid/phage vector and multiplied in *E. coli* or, amplified by PCR, separated on electrophoresis, purified from gel and cloned [9].

The most commonly synthesized genes range in size from 600 to 1,200 bp although, much longer that genes made by connecting previously assembled fragments of fewer than 1,000 bp. In this size range it is necessary to test several candidate clones confirming the sequence of the cloned synthetic gene by automated sequencing methods [23].

6. Cloning vectors

The molecular cloning brings the possibility to isolate, analyze, synthetize and clone individual genes or segments of DNA, creating a recombinant DNA. After isolated and purified the DNA target sequence must be mounted on an appropriate carrier molecule, the cloning vector.

A cloning vector is a small piece of DNA into which a foreign DNA is inserted for transfer or propagation in an organism, with the ability to self-replicate. The purpose of a vector is to allow efficient high-level expression of cloned genes or still, the need to increase the number of copies of a recombinant DNA [28].

6.1. Need to increase the number of copies of recombinant DNA

Besides having a DNA molecule already recombined, single copies are not sufficient to construct a recombinant DNA. The *in vitro* manipulation like, purification and transfer to a target cell, of a single copy is not possible. Thereby the recombinant construct should be propagated to increase the copy number. A convenient way to copy such fragments is to use the replication machinery of an organism, inserting the donor DNA in a cloning vector [29].

The essence of molecular cloning is to use restriction nucleases to cut DNA molecules in a starting DNA population (the target DNA) into pieces of manageable size, then attach them to a replicon (any sequence capable of independent DNA replication) and transfer the resulting hybrid molecules (recombinant DNA) into a suitable host cell which is then allowed to proliferate by cell division. Because the replicon can replicate inside the cell (often to high copy numbers) so does the attached target DNA, resulting in a form of cell-based DNA amplification [11].

6.2. Cloning vectors

In principle, any molecule of DNA that can replicate itself inside a cell system could work as a cloning vector, but many factors as, small sizes, mobility between cells, easy production and detection mechanism should be considered [28].

The type of host cells used in a particular application will depend mainly on the purpose of the cloning procedure. Host cells exploited are modified bacterial, fungal cells (*e.g.* Yeast), or still virus, being the bacterial system (*e.g. E. coli*) the most used due to their capacity for rapid cell division and for attend the major vectors requirements.

The vector may have an origin of replication that originates from either a natural extrachromosomal replicon or, in some cases, a chromosomal replicon [11]. Besides the structure the vectors should contain a sequence that make possible to select the recombinant cells, like a marker gene and in third place they should contain restriction sites into which the DNA can be inserted [29].

The types of cloning vectors are plasmids, phages, cosmids, phagemids, artificial chromosomes, viral vector and transposons. Each of them will be briefly describe in this section.

6.2.1. Plasmideal vectors

Plasmids are small circular double-stranded DNA molecules, which exist in the cell as extrachromosomal units. In a cell, they have the ability for self-replicating, and copy numbers maintenance. Due to their capacity of copy numbers they can be classified as: single copy plasmids or multicopy plasmids.

The single copy plasmids are maintained as one plasmid DNA per cell, instead the multicopy plasmids that are maintained as 10-20 copies per cell. Another kind of plasmids consists in ones that are under relaxed replication control, allowing their accumulation in numbers up to 1000 copies per cell, being the used ones as cloning vectors [15].

The plasmids vectors are designed to work in bacteria cells. An important property in these vectors is the detection of the same in the host cells. Usually, the detection mechanisms are done through antibiotic resistance. The host cell strain chosen is sensitive to a particular antibiotic and the plasmid is designed to contain a gene conferring resistance to this antibiotic.

Another approach for detection is through β-galactosidase gene complementation in which the host cells are mutants containing a β-galactosidase gene fragment and plasmid vector are designed to contain a different fragment of the same gene. By this way, after transformation functional complementation occurs and the host cells, which incorporate the plasmid are capable of β-galactosidase production.

The functional β-galactosidase activity can be accessed by conversion of a colorless substrate, Xgal (5-bromo, 4-chloro, 3-indolyl β –D- galactopyranoside) to a blue product [11]. The both methods are efficient for clone's selection, and their use depends on individual's preferences.

According to P. K. Gupta (2009) [15], there were three phases of plasmid development cloning vectors. The first included the plasmids pSC101, ColE1 and pCR1, which are naturally occurring plasmids, and not suitable for efficient cloning, since plasmid can transfer the gene through bacterial conjugation or can be integrated in the bacterial genome having no accessible detection system. Other disadvantage lies on having no more than two restriction sites for cloning.

The drawbacks of naturally occurring plasmids were overlapped by pBR313 and pBR322. pBR313 was too large having fifty percent of its sequences being non-essential. The size reduction brought the pBR322, which was largely used for many years.

The second phase relies on reducing the plasmids sizes, because the transformation efficiency and vector size have a proportional inverse relation. Thus, variations of the pBR322 appeared, including pAT153, pXf3, pBR327, etc. This plasmid vectors incorporate the selection mechanism of antibiotic resistance (described above).

The third phase involves incorporation of sequences for alpha-complementation selection (described above); incorporation of sequences from single strand M13 phage, for sequencing templates production; and, also integration of promoters' sequences, for *in vitro* transcription or expression of large amounts of foreign proteins. In this phase, plasmids like pUC, pGEM, M13, were developed.

Nowadays, there are a lot of plasmids commercially available that can be purchased depending on the application needs.

6.2.2. Lambda phage vectors

A bacteriophage lambda is a bacterial virus that infects *E. coli*. Its utility as a cloning vector depends on the fact that not all of the lambda genome is essential for its function [1]. The lambda genome has the left-hand region with essential genes for the structural proteins and the right-hand region has genes for replication and lysis, while the middle region has the genes for integration and recombination, which are non-essentials.

There are two possible types of lambda vectors: the insertion vector and the replacement vector.

The insertion vector has only a single recognition site for one or more restriction enzymes, enabling the DNA fragment to be inserted into the lambda genome. The lambda particle integrates DNA molecules between 37 and 52kb, and to adapt longer inserts is necessary to remove some of lambda genome. The region for replacement is the middle one where, more 23 kb of foreign DNA can be inserted. This vector is known as replacement vector [28].

The replacement vector cannot be integrated into the host cells chromosome being necessary to use a helper phage to provide integration and recombination functions. On the other hand, this vector has two restriction sites, having a whole section of phage genome being replaced during cloning [1].

6.2.3. M13 phage

M13 is filamentous bacteriophages that infect specific *E. coli*. Your attractive as a cloning vector consists in its genomes contain the desirable size for a potential vector (less than 10kb); does not kill the host when progeny virus particles are released and thus, is easily prepared from an infected *E. coli* cells culture. Besides, M13 is used as cloning vector to make single stranded DNA for sequencing and mutagenesis approaches.

The M13 genome is a single-stranded DNA molecule with 6407bp in length. This bacteriophage only infects bacteria carrying the F-pili (fragile protein appendages found on conjugation-proficient cells), being male-specific. When the DNA enters the cell, it is converted to a double-stranded molecule known as replicative form, which is a template for making about 100 copies of the genome. At this point replication becomes asymmetric, and single-stranded copies of the genome are produced and extruded as M13 particles. The property of do not lyse the host cell brings a DNA resource, although growth and division are slower than in non-infected cells [1, 11, 28].

6.2.4. Cosmids

Cosmids are plasmid particles into which certain specific DNA sequences, namely those for *cos* sites, are inserted. The goal of these vectors development is to cloning of large DNA fragments (up to 47kb in length). They are made up of plasmid sequences joined with lambda vectors sequences, trying to conjugate the properties of this both vectors in one (being transfected as a lambda vector by packaging/ infection mechanism and behaving as a plasmid when introduced into an *E. coli* cell).

The advantages consist of a highly efficient method of introducing the recombinant DNA and, a cloning capacity twofold greater than the best lambda replacement vectors. On the other hand, the gains of using cosmids instead of phage vectors are offset by losses in terms of ease to use and further processing of cloned sequences [1].

The methodology to use the cosmid cloning vectors consists in put together the cleaved vector and the target DNA for cloning, producing concatameric molecules. The concatameric molecules are usually generated by first linearizing the cosmid so that each end has *cos* site. Then the linear cosmid is cut with a *Bam*HI, which generates sticky ends with the overhang sequence GATC. The foreign DNA is also digested with *Mbol,* which also generates a GATC overhang. Partial digestion leaves some site uncut and allows large segments of a genome to be isolated. These segments are mixed with the two halves of cosmid and joined using ligase. Thus, these molecules are packaged into phage heads by mixing with a packaging extract, becoming infectious. *E. coli* cells are infected with the cosmids, and after infection the cosmid circularizes and multiply as a plasmid vector [15, 28].

6.2.5. Phagemids

Phagemids combine desirable features of both plasmids and bacteriophages. The construct consists of a plasmid with a segment of a filamentous bacteriophage, such as M13, having two different origins of replication: the plasmid and the phage origin. The selected phage sequences contain all the *cis*-acting elements required for DNA replication and assembly into phage particles [11, 30].

These vectors allow successful cloning of inserts several kilobases. After *E. coli* suitable strain transformation with a recombinant phagemid, the bacterial cells are superinfected with a filamentous helper phage, activating the phage origin and the phagemid. The plasmid DNA creates single stranded DNA, which is secreted into phage particles. These particles contain a mix of recombinant phagemids and helper phage. The selection is usually done by β-galactosidase gene complementation and by antibiotic resistance.

Vector pairs that have the phage origin in opposite directions are available, and as a result single stranded DNA representing of both DNA strands are produced. This mixed single strand DNA population can be used directly for DNA sequencing, if the primer for initiating DNA synthesis is designed to bind specifically to sequences of phagemid adjacent to the cloning site [11, 30].

Both cosmids and phagemids are characterized as hybrid vectors.

6.2.6. Chromosome Bacterial Artificial (BAC)

A bacterial artificial chromosome (BAC) is a single copy bacterial vector based on a functional fertility plasmid (F-plasmid) of *E. coli,* which can accept very long inserts of DNA between 300-350kb and allows the maintenance of many structural characteristics of the native genome.

BAC vectors are superior to other bacterial system, due to the F factor, which has genes regulating its own replication and controlling its copy number. These regulatory genes are *oriS* and *repE,* mediating unidirectional replication and *parA* and *parB,* maintaining the copy number to one or two per cell. The cloning segment includes the lambda bacteriophage *cosN* and the P1 *loxP* sites; two cloning sites (*HindIII* and *BamHI*); and, several C+G rich restriction enzyme sites (*Not I, Eag I, Xma I, Sma I, Bgl I* and *Sfi I*) for potencial excision of the inserts. The cloning site is flanked by T7 and SP6 promoters for generating RNA probes for chromosome walking and for DNA sequencing of the inserted segment at the vector-insert junction. The

CosN and loxP sites provides convenient generation of ends that can be used for restriction-site mapping to arrange the clones in an ordered way [31].

Besides the maintenance of large DNA inserts, BAC has structural stability in the host, high cloning efficiency and easy manipulation of cloned DNA, being largely utilized for construction of DNA libraries from complex genomes and subsequent rapid analysis of complex genome structure [31].

For recombination with DNA inserts, after enzymatic digestion DNA ligase are used. Transformed suitable *E. coli* was carried out by electroporation, and the competent cells are cultivated first with gentle shaking on liquid medium and then spreading to LB plates. The selection of recombined cells is done by hybridization procedures.

6.2.7. *Animal virus*

Viral vectors are commonly used to deliver genetic material into cells for gene therapy due to specialized molecular mechanisms to efficiently transport their genomes inside the cells they infect. This process can be performed inside a living organism (*in vivo*) or in cell culture (*in vitro*), being frequently used to increase the frequency of cells expressing the transduced gene [32].

The first use of vector virus for cloning was based on simian virus 40 (SV40), a polyomavirus originated of rhesus macaque, being a potent DNA tumor virus infecting many types of mammal cells in culture. The SV40 genome is 5.2 kb in size and contains genes coding for proteins involved in viral DNA replication, and genes coding for viral capsid proteins. Due to packing limitations, cloning with SV40 involves replacing the existing genes with the foreign DNA [32-34].

The other kinds of virus used for mammals' gene clones are adenoviruses, papillomaviruses, adeno-associated virus, herpes simplex virus (HSV), poxvirus and more recently retroviruses.

Adenoviruses came to solve the size of insert drawback of SV40, enabling the cloning of DNA fragments up to 8kb. On the other hand, due to its larger genome, adenoviruses are difficult to handle. Expression can be transient and the *in vivo* transfection can be impaired due to immune response.

Papillomaviruses also have a high capacity for inserted DNA with the advantage of stable transformed cell line.

Adeno-associated virus has this name because it is often found in cells that are simultaneously infected with adenovirus. To complete the replication cycle the adeno-associated virus uses proteins already synthesized by adenovirus, which acts like a helper virus. Lack of helper virus made the genome of adeno-associated virus integrate to host DNA. The major advantage of this vector consist of a defined the insertion site, always in the same position, being important in researches that cloning gene needs rigorously check such as gene therapy.

The herpesviruses include infections human viruses as herpes simplex virus (HSV), most used like a vector. The HSV is an enveloped double-stranded DNA, with 152kb, having advantages like larger foreign DNA carrying; high transduction efficiency and, potential to establish latency.

Poxvirus vectors are double-strand DNA with 200kb in the core and carrying up to 25kb o foreign DNA. Gene is stably integrated into the virus genome resulting in efficient replication and expression of biologically active molecules.

Many viruses kill their host cells by infection, so special artifices are needed if anything other than short-term transformation experiments is desirable. Bovine papillomavirus (BPV), which causes warts on cattle, is particularly attractive because they have an unusual infection cycle in mouse cells taking the form of a multicopy plasmid with about 100 molecules present per cell. This infection does not bring the death of cell and, BPV molecules are passed to daughter cells during mitosis.

The most used viral vectors are the retroviruses, infectious viruses that can integrate into transduced cells with high frequency, inserting the foreign DNA at random positions but, with great stability. They can be replicated-competent or replication-defective.

Replication-competent viral vectors contain all necessary genes for virion synthesis, and continue to propagate themselves once infection occurs. These vectors can integrate an inserted about 8–10 kb, limiting the introduction of many genomic sequences. This made replication-defective vectors the usual choice. These vectors had the coding regions replaced with other genes, or deleted. These viruses are capable of infecting their target cells but they fail to continue the typical lytic pathway that leads to cell lysis and death.

The viral genome in the form of RNA is reverse-transcribed when the virus enters the cell to produce DNA, which is then inserted into the genome at a random position by the viral integrase enzyme. The vector, now called provirus, remains in the genome and is passed on to the progeny of the cell when it divides. The site of integration is unpredictable, which can pose a problem; therefore, the principal drawback of retrovirus vectors involves the requirement for cells to be actively dividing for transduction, being widely used in stem cells. Great examples to overcome this disadvantage are lentiviruses vectors.

The lentivirus is a subset of retrovirus with the ability to integrate into host chromosomes and to infect non-dividing cells. Lentivirus vector systems can include viruses of non-human origin (feline immunodeficiency virus, equine infectious anemia virus) as well as human viruses (HIV). And for safety reasons lentiviral vectors never carry the genes required for their replication, preventing the occurrence of a wildtype-potentially infectious virus [32-34].

6.2.8. Transposons

DNA transposons elements are natural genetic elements residing in the genome as repetitive sequences that move through a direct cut-and-paste mechanism. This process is independent of previously recognized mechanisms for the integration of DNA molecules and occurs without need of DNA sequence homology. Thus, they can be used as tools from transgenesis to functional genomics and gene therapy.

Transposons are organized by terminal inverted repeats (ITRs) embracing a gene encoding transposase necessary for relocation. Transposons move through a "cut-and-paste" mechanism, known as transposition, which involves excision from the DNA and subsequent integration into a new sequence environment [35, 36].

The development of transposable vectors is based on a plasmid system, with a helper plasmid (expressing the transposase) and a donor plasmid (with terminal repeat sequences embracing the foreign gene) [36].

6.3. Importance of promoters

The promoters are defined as *cis*-regulatory elements responsible for the control of transcriptional machinery and determination of its level and specificity, marking the point at which transcription of the gene should start, and regulating the transcription. Promoters contain proximal elements, involved in the formation of the transcription complex; and, major elements that give cell specificity of protein expression [37, 38].

For long term transgenic expression *in vivo* or tissue specific expression, the transcription of the foreign gene should be controlled for promoters, which in this case are inserted on cloning vectors [37].

Approaches requiring a high ubiquitous expression of the transgene can be accomplished with non-tissue specific promoters. These promoters are actives in almost all of cell types, ensuring the foreign gene expression in all organism tissues. Examples of these promoters are metallothionein gene promoter, EF1 gene promoter, CMV early gene promoter, human H2K gene promoter, 3-methylglutaryl CoA reductase gene promoters, and others.

On the other hand, to restrict transgene expression to the target tissue the promoters used are tissue-specific. These promoters can direct the transgene expression to lung, epithelia, liver (albumin gene promoter), pancreas (amylase promoter), muscles (truncated muscle creatine kinase - MCK), neural cells (synapsin 1), mammary gland and cardiac cells (troponin T promoter), and so on [38]. Promoters used in cloning vectors should be sufficiently short to be cloned in a gene transfer vector.

Besides the use of tissue-specific promoters, another kinds of promoters are the inducible ones, which transcription can be selectively activated. These promoters respond to specific transcriptional activators are: transcriptional activators regulated by small molecules; intracellular steroid hormone receptors; and, synthetic transcription factors in which dimerization is controlled by antibiotics.

The promoters for transcriptional activators regulated by small molecules are based on the use of transcription factors that change their conformation upon binding one small chemical molecule (*e.g.* Tet repressor – TetR). The promoters for intracellular steroid hormone receptors act when hormone analogs are ligated to the hormone's modified receptors. Synthetic transcription factors in which dimerization are ones that in the presence of antibiotics tethers the transcriptional activation [37, 38].

7. Practical application of genetic Engineering and cloning: From transgenic animal models until cloning animal

Transgenic animal technology and the ability to introduce functional genes into animals are powerful and dynamic tools of genetic engineering. The genetic engineering field allows stable introduction of exogenous genetic information into any live organism, enhancing existing or, introducing entirely novel characteristics. The cloning technology is closely related with transgenic, being used as a tool for genetic engineering of an animal.

Together these technologies can be used to dissect complex biological process, like *in vivo* study of gene function during development, organogenesis, aging, gene therapy, and epigenetics studies. Besides, there are a lot of commercial applications like, model for human diseases, pharmaceutical biotechnologies development, and reproduction of a valuable animal.

7.1. A sheep named Dolly: Cloning

In 1997, Wilmut and coworkers announced Dolly production, which was the first mammal cloned from adult cells. In this experiment Dolly was born after reconstruction of 277 embryos with mammary gland cells.

Her birth at 5 July 1996 in Scotland brought huge excitement of the scientific world, beginning a biological revolution. The fact of Dolly has been created from adult differentiated cells showed the possibility not imagined before: dedifferentiation of already committed somatic cells, which brings a lot of repercussion. After Dolly, the differentiated cells cloning was achieved in a lot of species like, bovines, murines, caprines, swines, felines and canines [39-46]

7.2. What is cloning?

The definition of clone consists in the reproduction of genetically identical organisms, naturally or artificially, by asexual reproduction (without spermatozoa). The word "clone" comes from the greek word "klon", that means twig. With these characteristics clone for some organisms is a physiological asexual way of reproduction (e.g. bacteria and yeast). This conception, after Dolly's production went further, becoming the production of genetically identical live organism through nuclear transfer techniques. This defines clone to a process in which cellular material from a DNA donor is transferred to an egg whose own DNA has been removed, resulting after some procedures in embryo genetically identical to DNA of original cell clone.

The origins of nuclear transfer remount discoveries with amphibians by Spemann (1938), who demonstrated that nuclei of newt salamanders are pluripotent up to eight-cell stage, leading intensive studies with nuclear transfer in *Rana pipiens* and *Xenopus laevis*, attempting to understand the nuclei participation of differentiated cells in reprogramming. Studies by Brings and King (1952) showed that amphibian oocytes receiving blastula nuclei could be reared to maturity [47].

During 60 and 70 decades, nuclear transfer was done mostly in amphibian, leading to clones production from intestinal larvae cells, being the first evidence that differentiated cells keep the potential to form all tissues of an organism. [48].

In mammals the first nuclear transfer studies were done in mice, in which Illmensee and Hoppe (1981) reported that this technique could be used to produce mice clones from embryo cells. In domestic animals, Willadsen (1986) published the first report with lamb clones production. This accomplish was confirmed after with bovines, rabbits, swine, and others. And, at 1996 Dolly brought the accomplishment of mammal nuclear transfer form adult cells [48].

Nowadays, a lot of cloned animals could be produced, which besides the commercial interest of reproducing some valuable animal, made the technique used for research like reprogramming mechanism and epigenetics controls.

7.3. Producing a clone: Technical

Technically to produce a clone from nuclear transfer the majority of protocols are based on these steps: preparation of cytoplasm receptor; oocyte enucleation; preparation of nuclei donor cells; embryo reconstruction; artificial activation; embryos culture; and, embryo transfer. Each of them will be briefly described below.

7.3.1. Methodology, advantages and disadvantages

Initially to produce a clone animal by nuclear transfer is necessary to prepare the receptor cytoplasm. The receptor cytoplasm is a cell which nuclei was removed by in a process known as enucleation, being the most used cell the female egg: the oocyte. The oocyte can be used in a lot of division estate being, actually used in metaphase II.

To obtain the oocyte at metaphase II, first, or they are aspirated from ovaries from slaughterhouses or by Ovum Pick Up, ultrasound guided (used for domestic animals), being obtained before metaphase II, needing *in vitro* maturation or, being collected already at this estate, after *in vivo* maturation (mostly used for laboratory animals).

When maturation is needed, the oocytes are recovered from antral follicles at prophase I estate or germinate vesicle, and are maturated in temperature, medium and time specific, inside incubators, having the control of CO_2 tension. The two kinds of procedure to obtain a metaphase II oocyte, have its advantages and disadvantages. The *in vivo* maturation, are capable of better quality oocyte production, but depending on the species, it cannot be achieved, needing the *in vitro* maturation. The *in vitro* maturation besides not be the natural way of reproduction, brings good and quality results too, being largely used.

The second step consists of oocyte enucleation. Besides a lot of methodologies have been developed, the most used way to remove the nuclei from the oocyte is micromanipulation procedures.

Before, to prepare the oocytes for enucleation, the cumulus cells from the maturated ones are removed, which can be done mechanically by a tube agitator (vortex); with pipettes sized as thin as the oocyte; or, chemically by hialuronidase. After, the oocytes are carefully selected, based on the presence of the first polar body (checking if the metaphase II was achieved), and cytoplasm morphology.

The micromanipulation procedure for enucleation involves first fixing the mature oocyte by a holding pipete. To remove the nuclei, the first polar body is used to reference of the metaphysary plate (nuclei). At metaphase II the chromatin remains at the oocyte periphery, close to the polar body. The enucleation is done by an enucleation pipete with a bevel-shaped tip, which penetrates the pellucid zone (PZ) aspirating the first polar body e part of the cytoplasm attached to this structure.

Another enucleation method is the oocyte bisection [49]. In this case the PZ is removed and the oocytes are sectioned by a micro-blade in two halves, being removed 50% of the cytoplasm. The half with the nuclei is discarded and the other one used for nuclear transfer.

To check the efficiency of enucleation, the enucleated oocytes can be stained with DNA fluorescent dyes using in most of the cases Hoechst 33342 (H342), which need exposure of ultra-violet (UV) light to be verified.

This procedure depending on the time of exposure can compromise the oocytes viability. Trying to minimize this effect the exposure only of the removed cytoplasm and polar body to UV light can be done. The presence of chromatin in this material indicates the success of enucleation. There are protocols not involving fluorochromes like the use of demecolcin, incubating the oocytes 1-2 hours, in a medium containing this substance. This procedure creates a protrusion at oocyte membrane where the chromatin is localized [48].

The amount of cytoplasm removed at enucleation process direct interferes in embryos development rates. The less amount of cytoplasm removed, better rates of embryo development are achieved. Usually, the enucleation for bovines has an efficiency rate between 50-70% [46].

The third step is preparation of nuclei donor cells, which depends of the cell type and the technique used for nuclei transfer. The donor cells can be originated from embryonic, fetal or adult cells. When using embryonic cells, the PZ of embryos are removed by enzymatic digestion, acid solution or mechanically. The embryonic mass should be held on a calcium and magnesium free solution, facilitating the blastomeres disintegration. If the donor cell was fetal or adult mostly fibroblasts are used due to easy culture. A primary culture is done by a biopsy from a skin fragment. The cells are held on culture until the third passage, at least, due to homogeneity and specific cells reaching on the culture, more than the third passage cells can be used as nuclei donor too.

Besides a lot of experiments have been realized to determine which somatic cell type would be the most appropriated for cloning, until now is not yet known if some kind of cell are most advantageous for nuclear transfer [48].

The fourth step is embryo reconstruction consisting of place the nuclei from the donor cell inside the enucleated oocyte. This can be achieved by microinjection or membranes fusion, whereas the first has low results [46]. Using the fusion method with micromanipulators help, each cell is introduced at the perivitelline space of the enucleated oocyte. Then the fusion can be done by electric pulses (electrofusion), liposomes, polyethylene glycol, or still, by inactivated viruses. The electrofusion are the mostly used. In this method, the complexes receptor-donor nuclei are positioned at electrofusion chamber, where they are submitted to two electric pulses with low conductance, preventing heat dispersion. These pulses induce the membrane fusion incorporating the cell donor nuclei at the receptor cytoplasm.

The fifth step consists of artificial activation, which involves degradation of enzymatic complexes responsible for oocytes kept at metaphase II, being needed for accomplish the meiotic process initiating the embryonic development. At physiologic conditions, this is achieved with spermatozoa. But at cloning process, without the spermatic cell, chemically or physical methods are used (ethanol, electric pulses, calcium ionophore and, strontium chloride).

The activation moment of oocyte in relation with the nuclear transfer moment have important consequences at the chromatin integration and remodeling; viability and, embryo development [48].

The sixth step consists of embryo culture, in which the reconstructed and activated embryos are cultivated at CO_2 incubators, until the blastocyst stage (species time dependent). The culture conditions are similar with *in vitro* fertilization conditions, whereas cloned embryos are more sensitive to cryopreservation, and do not pass through expansion phase when blastocyst stage are achieved. The PZ rupture, done by enucleation process, made the expansion estate coincide with hatch estate, and at this estate the embryos are transferred to synchronized female receptors, and after gestation and parturition or caesarean, cloned animals are produced [46].

7.4. What is a transgenic animal?

A transgenic animal consists of an animal whose genetic material has been altered using genetic engineering techniques. Foreign DNA is introduced into the animal, using recombinant DNA technology, and then must be transmitted through the germ line so that every cell, including germ cells, of the animal contains the same modified genetic material [32, 50].

S. N. Cohen and H. Boyer generated a functional organism that combined and replicated genetic information from different species, creating the first genetic modified organism in 1973. In 1974 R. Jaenisch created the first genetically modified animal by inserting a DNA virus into a mouse embryo showing inserted genes was present in every cell. However the mice did not transmit the transgene. In 1981 F. Ruddle, F. Constantini and E. Lacy injected purified DNA into a single-cell mouse embryo and showed transmission to subsequent generations. During the early eighties the technology used to generate genetically modified mice was improved into a tractable and reproducible method [51, 52].

7.5. Producing a transgenic: Technical

Currently, the three most widely used procedures for creating transgenic animals are microinjection of the cloned gene(s) into the pronucleus of a fertilized egg, injection of recombinant embryonic stem cells into embryos, and the use of retroviruses. There are other methods like sperm cells mediated gene transfer; *in vivo* gene transfer and ICSI-transgenes. These methods will be briefly discussed below.

7.5.1. Methodology, advantages and disadvantages

The microinjection of foreign DNA directly into the pronuclei of fertilized zygotes is the most extensively and successfully used method of gene transfer in the mouse. This method was the first non-viral method for transgenic animal production. The DNA microinjection to pronucleus has low technical progress, but was disseminated for other species (rabbit, swine and goats) [53, 54].

To produce a transgenic animal a lot of zygotes are needed which is achieved by female superovulation before mating. For mouse, rats and rabbits the one cell embryos are transparent, being opaque in swine, goats, sheeps, and cow, due to lipid presence. In case of opaque embryos they should be centrifuge before the microinjection for concentration of lipids at one embryo side, allowing the pronuclei visualization.

The disadvantages of this technique are due to exogenous DNA introduced at the pronucleus is strongly mitogenic, leading a lot of embryos microinjected to death. Another disadvantage consists of integration of foreign DNA in a random manner, being not possible to predict the integration site and control the number of copies of transgenic integrated DNA. The transgenic production by microinjection to pronucleus are 2% for mouse, 0,1-0,5% for pigs, 0,01-0,1% for sheeps and goats and lower for cows [54].

Another used methodology is infection by retroviruses, in which the transgenes can be introduced by viral infection of preimplantation embryos. Retroviruses have the natural ability to infect cells and integrate its genome to infected cells. The retroviruses are modified, in which some of its genes are substituted by a target gene. Then these reconstructed viruses are transferred for cells, that after infected synthesize viral protein, secreting viral particles which can infect embryonic cells, or primordial germ cells, originating transgenic animals.

The disadvantages came from biosecurity, since this technique works with recombined viruses. This determines biosecurity rules to be carefully followed, preventing this vector dissemination. One advantage is the use of this method is for gene therapy, transferring the modified viruses for somatic cells of the patients. And another advantage came with lentiviruses uses that are successfully used for gene transfer (*see item 5 for advantages and disadvantages of this virus*).

The gene transfer by injection of recombinant embryonic stem cells into embryos is one of the most useful when is necessary to select for rare integration events or when is necessary an chimeric animal production [53].

Embryonic stem cells (EST) are cell lineages obtained from initial embryos estate, like morula or blastocyst. These cells have pluripotency capacity, being capable of participation actively of all organism tissue production, including the gametes. The technique for gene transfer consists in transfect EST with exogenous DNA; and the ones transfected after been selected, are introduced into embryos. The resulting animals are chimeric and mosaics for the transgene, and if the germ cells have integrated the transgene, this can be passed for the progeny.

The disadvantages of this methodology are the impossibility of chimeric animals from other species besides mouse; transmit the transgene to their progeny. Being an advantage when the transgene requires homolog recombination.

Another ways of transgenic animals production are being used like, the use of gametes cells to transgenic animal production are already achieved. The oocytes uses do not generated enough transgenic animals, being the spermatozoon most appropriated cells for transgenesis [54].

The Sperm-mediated gene transfer (SMGT) enables the production of transgenic animals by exploiting the ability of sperm cells to bind and internalize exogenous DNA. The SMGT has being an easy and low cost method for transgenic animals production, in which the simple incubation of the exogenous DNA with the sperm, follow by artificial insemination or IVF procedures, can result in an transgenic animal. Although, a lot of transgenic animals produced by this technique, there are high number of studies with low reproducibility and the reason for this are unknown.

To pass through this low reproducibility the spermatic cell are been used for integration of the new genetic material by intracytoplasmic sperm injection (ICSI-transgenesis), method largely and successfully used for mammalian species other than mouse [53].

The *in vivo* transfer of a transgene can be done by the injection of the exogenous DNA to testicles or still blood veins, bringing good results. The revolutionary of this method came with production of *in vitro* sperm stem cell, from many species. After cultivated this primordial sperm cells are transfected with the exogenous DNA, then the recipient animal are treated for decrease its physiological sperm production, and the transfected cells are injected trough testicle efferent duct. Thus, after the spermatogenesis this male can be used for transmission of transgenic sperms.

7.6. Transgenic animals such as experimental models

Genetically modified animals currently being developed can be different broad intended purpose of the genetic modification: to improve animal production; xenotransplantation; to produce proteins intended for human therapeutic use; to improve animals' interactions with humans trough hypo-allergenic pets; to improve animal health by disease resistance animals; and, to research human diseases with the development of animal models [50].

The animal models production are used in pathologies or syndromes caused by inactivation or dysfunction of a determinate gene, being possible to delineate strategies and prepare the equivalent genetic modification in a homologous gene of the animal [55].

The use of animal models to research human disease, are done because of similarity to humans genetics, anatomy, and physiology or for being easier to have a lot of conditions developed in many transgenic animals, manipulating just one variable each time, which corroborate with statistical analyses of the results.

Extensive research for human diseases have been done with rats, mice, gerbils, guinea pigs, and hamsters, being the mice the mostly used due to genomic similarities to human and easy and developed handle and production methods; low cost; and, high reproductive rates.

Although these advantages sometimes the small size of mice leads to challenges in the design and application of instrumentation for physiological measurements, being possible to introduce transgenes into larger mammals and also fish like genetically modified zebrafish. The advantages using model experimental organisms larger than mice are easier physiological assessments and besides, provide alternatives when manipulation of the mouse genome does not produce the phenotype one wishes to investigate. Examples of this situation are provided by the hypertensive response of rats but not mice to forced expression of the *REN-2* gene and the more severe spondyloarthropathy produced by *B27* and b2-microglobulin transgenes in rats [56].

Multiple models of diverse pathologies have been generated, like diabetes, obesity, allergy, cancer, cardiovascular disorders, hypertension, embryo development abnormalities and, reproductive system abnormalities [57].

Sometimes, complex animal models are needed, using coexistent modification of genes trough techniques like knockout, knock in and knock down. Another approach is the use of news genetic modification variants, like inducible expression of transgenes and restriction of transgene expression at some organs. These techniques can generate animal models that better fit some human pathology [57].

7.7. How to join cloning with genetic modification: Complementary biotechnologies

The cloning technology together with genetic modification of organism originates a new method for animal transgenesis production. Between different areas of application of nuclear transfer, the transgenesis has the major benefits with its advances. Besides laborious, this technique allowed more efficiency transgenic production than pronuclear microinjection for ruminant animals.

The advantages of this methodology for animal transgenesis production are introduce, functionally delete or subtly genes of interest; produce embryos expressing the transgene constitutively not mosaic or chimeric; or still, achieve genetic modifications directed. This last approach can be done by insertion of genes in a determined chromosome position.

The genes inactivation by the cloning technique for transgenic production, are achieved by substitution from homolog recombination, which largely obtained just in somatic cells. Examples of this approach in pigs was the substitution of β-galactosiltransferase, making its kidneys being resistance to hyperacute rejection in to experimental transplant in non-human primates [56].

Another importance of this approach consists of the applications of transgenic farm animals. The major opportunity of this use is biopharming, mostly produced at the milk of transgenic female animals, in which cells modified containing important human genes under a promoter for mammary gland control (*for more information of promoters see topic 5*). After milk secretion this therapeutic protein are purified and used for clinical trials to evaluate their safety and effectiveness treating human diseases and disorders before gaining regulatory approval. Other tissues examples, used for antibodies production are eggs of chicken or blood of transgenic cattle [58].

8. Future Prospects — Genetic engineering and cloning: A dream or a nightmare?

8.1. Emphasis on cloning technology

A clone is an identical copy of the parental material, which is development. Clones from a same cell type will have the same genomic properties, but in multicellular organisms different cell behavior and phenotypic are observed by the influence of environment. Cloning is not in itself a genetic engineering technique as transgenic but both techniques are strongly correlated. Cloning naturally may occurs after a single insemination and if a specific gene is not inserted into the host genome the resulting animal cannot be considered a genetically modified organism, transgenic [59]

It is well established in literature that clone embryos have a lower total number of cells when compared to embryos not cloned. Cloned bovine embryos have approximately 9% less cell, and this rate is 19, 43 and 55% charge respectively for pigs, rabbits and mice cloned embryos and this rate being positively correlated with the difficulty of performing cloning [60]. The discovery of cell types that offer better rates cloning is essential to the development and better understanding of the epigenetic molecular mechanisms, nuclear programming and reprogramming; thus, developing more secure and efficient cloning techniques [61]. Use races which have highest cloning success rates as a model, Ex. Japanese Black cattle, could help to find the key points to solve currently problems as high mortality rate of cloned offspring. In 2003, a study compared the size of spermatozoon chromosome telomeres obtained from cloned and not cloned animals. For both groups was observed that telomere length was maintained throughout animals age, fact which indicated that cloned animals could be used as breeders [62].

After Dolly´s birth, one glimpses in human medicine to use cloning technologies in gene therapy for tissues and organ replacement without risk of transplant rejection, since the donor would be the patient himself. For some people, the ability to clone differentiated cell is regarded as the long awaited immortality or as an insult to religious principles. Nowadays cloning assumes a prominent role in the world media among the most controversial issues. However, if would possible to produce healthy cloned offspring with low mortality and without genotypic and phenotypic changes, why would clone be so controversial? It is important remember that similar process naturally occurs, once the medical literature shows that one in 250 births produces the identical twins, the current human clone.

Currently cloning technology is in emphasis on science and media, and over the last few years has made important advances. However, more knowledge is needed to this technique able to exercise your full potential as a biotechnology in combination with gene therapy and engineering gene.

8.2. Ethical aspects of genetic engineering: Risks and benefits

The generation of the genetic engineering and his major advances in biotechnology as: sequence of complete genome for different plants and animals, creation of transgenic and cloning had the participation of different scientists.

Mendel is considered the father of modern genetics. Their findings were published in 1865 and just after 50 years (1909) were considered by Wilhelm Johannsen who discovered the genes, which were called before as "elements" by Mendel. In 1953, James Watson and Francis Crick published in Nature the discovery of the helical structure of DNA. At 1973, the concept of genetic engineering was created by S. Cohen and H. Boyer who published the ability to "cut and paste" the genetic material. However, it was Van Rensselaer Potter the responsible for the term Bioethics through his book Bioethics: bridge to the future (1971), which was an answer for the numerous innovations in science. The advances in this field culminated in the announcement of the complete sequencing of the human genome, being followed by other species sequencing.

The discoveries and advances of genetic engineering brought an ethical dilemma. For large population, this science has turned into a dangerous knowledge because we are accumulating information faster than ability to manage them. This creates a conflict between ethical principles and moral norms. Ethics is the part of philosophy that studies the moral ideals and principles for human behavior. However, morality is grounded in obedience to customs and habits. Ethics is different from moral because it is based on moral actions (good and bad) generated by reason. Thus, the concept of bioethics is the applicability of ethics in biological science, its discoveries and advances.

Using these definitions, genetic engineering is not in its bioethical concept harmful to society. Fear of the unknown by lack of information and knowledge has made us to come back at ancestors response. The modern society create new myths and rites. However, with technological advances in genetic engineering the natural boundaries were lost. Now it is the duty of man to establish and determine these new barriers. Genetic engineering should be rather limited and controlled by new ethical and technical barriers.

With industrial development and new technologies we are living in a new era, the biotechnology era. With advances, especially in genetics, a new science called genetic engineering that enables the manipulation (insertion and removal) of genetic material was created. This technique can be applied since one cell organism (bacteria) until complex multicellular organisms such as farm animals and humans. The genetic engineering, as the industry, searches for new products (life forms) that have greater production efficiency.

The welfare and environmental concerns for genetic engineering were very well discussed by Fox, MW (1998)[63]. According to the author, bio processing industry proposes the production of new forms of energy, synthetic and pharmaceuticals products. However, many promises of innovation are accompanied by profound discussions regarding the risks and ethical concepts. These aspects should not only be discussed by governments and industries, but also for the entire population and researchers. Generally genetic manipulation of bacteria and plants, which are transformed into real machines for the production of biological products such as hormones, is considered moral and ethical acceptable. This is explained because plants and bacteria not have the ability to Suffer, to experience pain and emotional distress. Thus, would it be ethically acceptable to use more complex organisms, such as farm animals, as biological models and bioreactors? Had genetic manipulation any effect on your body structure and physiology? If any error occurs during the process, is the birth of animals with anomalies acceptable?

A study conducted in the UK in 1997 by Frewer et al.[64] consider public concepts on application of specific and general genetic engineering. The main concepts related to rejection of genetic engineering were: personal objections, immoral, unnatural, unethical, harmful, personal worry, negative welfare effects, dangerous risk, tampering with nature and creation of inequalities. The positive aspects were: beneficial, advantageous, necessary and progressive. The data closely relate applicability of genetic engineering with risks and benefits that are defined by the nature of their application. The activities considered more negative were associated with genetic manipulation of animals and humans, confirming results previously obtained by previous researchers. The most important for the establishment of a concept is the applicability of the technology, most humans and animals that plants and microorganisms. The results imply that the public attitudes are defined by the process associated with genetic engineering rather the product of this process. Unnaturalness is one of the most important concepts associated with animal and human genetic material. The medical risk is high in benefice and low in risk and it is considered acceptable. However, non-medical applications are low in benefice and high in risk and it is unacceptable.

The complexity of public concepts about genetic engineering generated a hierarchical model for the dissemination of information that was proposed in 1990 by Hilgartner [65], which uses concepts of public debates to enter public opinion. Thus, the ethical and moral concepts of society are fundamental to be associated at genetic engineering to it exerts your complete function and optimal use. The public ethical concept must be considered particularly to genetic engineering involving animals and humans.

Moreover, models of diffusion of knowledge about risks and benefits of genetic engineering to society should be applied. Only with more information and more knowledge you can get over rampant enthusiasm and ideological fear. So some paradigms about genetic manipulation will be clarified. The biggest problem in genetic is not to make changes in the DNA, but to be faithful to a principle which are common to all men, of all cultures and responsible for perpetuating human and natural environment; therefore more important than any gene is ethics [66]. So, to finish this chapter there is no better phrase than: ethics is responsible to regulate and coordinate the genetic engineering, as also occurs in the others branchs of science.

Acknowledgements

FAPESP, CNPq and CAPES

http://dx.doi.org/10.5772/56071

Author details

Mariana Ianello Giassetti*, Fernanda Sevciuc Maria*,
Mayra Elena Ortiz D'Ávila Assumpção and José Antônio Visintin

Laboratories of *in vitro* fertilization, cloning and Animal Trasngenesis, Department of Animal Sciences, Veterinary Scholl, University of Sao Paulo – Sao Paulo, Brazil

*The first two authors contributed equally

References

[1] Nicholl DST. An Introduction to Genetic Engineering: Cambridge University Press; 2008.

[2] Lewin B. Genes 8: Pearson Prentice Hall; 2004.

[3] Satya P. Genomics and Genetic Engineering: New India Publishing Agency; 2007.

[4] Brown TA. Genomes: Wiley-Liss; 2002.

[5] Shapiro, J. L.; MacHattie, L. E.; Ippen,G. I. K.; Beckwith, J.; Arditti, R.; Reznikoff, W.; Mac-Gillivray, R. The isolation of pure lac operon DNA. Nature, 1969, 224:768-774.

[6] Korman, A. J. Knudsen, P. J; Kaufman, J. F.; Strominger, J. L. cDNA clones for the heavy chain of HLA-DR antigens obtained after immunopurification of polysomes by monoclonal antibody. Proceedings of National Academy of Science of United States of America.1982, 79:1844-1848.

[7] Maxam, A. M.; Gilbert, W. A new method for sequencing DNA. Proc. Nati. Acad. Sci. USA, 1977, 74 (2): 560-564.

[8] Graham CA, Hill AJM. DNA Sequencing Protocols: Humana Press; 2001.

[9] K GP, Gupta PPK. Biotechnology And Genomics: Rastogi Publications; 2004.

[10] Perkin Elmer. Automated DNA Sequencing Chemistry Guide. Perkin Elmer Applied Biosystems. 1998, 3-22, 3-27

[11] Strachan T. Human molecular genetics: Taylor & Francis; 2003.

[12] Shinwari J. Automated DNA sequencing. Megabace 1000, 2005.

[13] Kocher TD. PCR, directing sequencing and the comparative approach. PCR methods and applications. 1992:217.

[14] Rao V. Genome Res. 1994;4:S15-S23.

[15] Gupta PPK. Molecular Biology and Genetic Engineering: Rajpal And Sons Publishing; 2008.

[16] Pease , A.C.; Solas, D.; Sullivan, E. J.; Cronin, M. T.; Holmes, C. P.; Fodor S.P. A. Light-generated oligonucleotide arrays for rapid DNA sequence analysis. 1994; 91: 5022-5026.

[17] Stears RL, Martinsky T, Schena M. Trends in microarray analysis. Nat Med. 2003 Jan; 9(1):140-5.

[18] Chaudhuri, J.D. Genes arrayed out for you: the amazing world of microarrays. Medical Science Monitor, 2005; 11(2):52-62.

[19] Rosa GJdM, da Rocha LB, Furlan LR. Estudos de expressão gênica utilizando-se microarrays: delineamento, análise, e aplicações na pesquisa zootécnica. Revista Brasileira de Zootecnia. 2007;36:185-209.

[20] Jurinke C, Oeth P, van den Boom D. MALDI-TOF mass spectrometry: a versatile tool for high-performance DNA analysis. Mol Biotechnol. 2004 Feb;26(2):147-64.

[21] Gut IG. DNA analysis by MALDI-TOF mass spectrometry. Hum Mutat. 2004 May; 23(5):437-41.

[22] Mauger F, Bauer K, Calloway CD, Semhoun J, Nishimoto T, Myers TW, et al. DNA sequencing by MALDI-TOF MS using alkali cleavage of RNA/DNA chimeras. Nucleic Acids Res. 2007;35(8):e62.

[23] Artificial gene syntheis http://en.wikipedia.org/wiki/Artificial_gene_synthesis#cite_note-Edge81-2 (accessed 1 September of 2012).

[24] Temin HM, Mizutani S. RNA-dependent DNA polymerase in virions of Rous sarcoma virus. Nature. 1970 Jun;226(5252):1211-3. PubMed PMID: 4316301. eng.

[25] Sambrook J, Russell DW. Molecular Cloning: A Laboratory Manual: Cold Spring Harbor Laboratory Press; 2001.

[26] Stemmer WP, Crameri A, Ha KD, Brennan TM, Heyneker HL. Single-step assembly of a gene and entire plasmid from large numbers of oligodeoxyribonucleotides. Gene. 1995 Oct;164(1):49-53.

[27] Young L, Dong Q. Two-step total gene synthesis method. Nucleic Acids Res. 2004;32(7):e59.

[28] Clark DP, Pazdernik NJ. Molecular Biology: Academic Press; 2012.

[29] Pratik, S. Genomics and Genetic Engineering New India Publishing, 2007.

[30] Corley RB. A Guide to Methods in the Biomedical Sciences: Springer; 2004.

[31] Shizuya H, Birren B, Kim UJ, Mancino V, Slepak T, Tachiiri Y, et al. Cloning and stable maintenance of 300-kilobase-pair fragments of human DNA in Escherichia coli using an F-factor-based vector. Proc Natl Acad Sci U S A. 1992 Sep;89(18):8794-7.

[32] BUY hpucbth. BUY, http://people.ucalgary.ca/~browder/transgenic.html. (accessed 3 September of 2012.

[33] Working with viral verctor http://www.stanford.edu/dept/EHS/prod/researchlab/bio/docs/Working_with_Viral_Vectors.pdf (accessed 3 September of 2012.

[34] Gene Clonning, chapter 7. http://www.blackwellpublishing.com/genecloning/pdfs/chapter7.pdf. (accessed 3 September of 2012.

[35] Grabundzija I, Irgang M, Mátés L, Belay E, Matrai J, Gogol-Döring A, et al. Comparative analysis of transposable element vector systems in human cells. Mol Ther. 2010 Jun;18(6):1200-9.

[36] Meir YJ, Wu SC. Transposon-based vector systems for gene therapy clinical trials: challenges and considerations. Chang Gung Med J. 2011 2011 Nov-Dec;34(6):565-79. 37.

[37] Giaca, M. Gene Therapy, Springer, 2010.

[38] Pinkert, C.A., Transgenic Animal Technology. A laboratory handbook. 2nd Ed. Academic press, 2002.

[39] Mello, M.R.B. ; Caetano, H. V. A.; Marques, M. G.; Padilha, M. S.; Garcia, J. F.; Assumpção, M.E.O.A.; Lima, A. S.; Nicácio, A. C.; Mendes, C. M.; Oliveira, V. P.; Visintin, J. A. Production of cloned calf from fetal fibroblast cell line. Brazilian Journal of Medical and Biological Research, 2003, 36: 1485-1489.

[40] Kato, Y.; Tani, Y.; Sotomaru, Y.; Kurokawa, K.; Kato, J.; Doguchi, H.; Yasue, H.; Tsunoda, Y. Eight calves cloned from somatic cells of a single adult. Science,1998, 282: 2095-2098.

[41] Onishi, A.; Iwamoto, M.; Akita, T.; Mikawa, S.; Takeda, K.; Awata, T.; Hanada, H.; Perry, A.C. Pig cloning by microinjection of fetal fibroblast nuclei. Science, 2000, 289: 1188-1190.

[42] Shin, T.; Kraemer, D.; Pryor, J.; Rugila, J.; Howe, L.; Buck, S.; Muphy, K.; Lyons, L.; Westhusin, M. A cat cloned by nuclear transplantation. Nature, 2002, 415: 6874-6859

[43] Wakayama, T.; Perry, A.C.; Zuccotti, M.; Johnson, K.R.; Yanagimachi, R. Full-term development of mice from enucleated oocytes injected with cumulus cell nuclei. Nature, 1998, 23:369-74.

[44] Woods, G. L.; White, K. L.; Vanderwall, D. K.; Li, G.; Aston, K. I.; Bunch, T. D.; Meerdo, L. N.; Pate, B. J. A Mule Cloned from Fetal Cells by Nuclear Transfer. Science, 2003, 22: 1063.

[45] Lee, B.C.; Kim, M.K.; Jang, G. O. H. J.; Yuda, F.; Kim, H.J.; Shamim, M.H.; Kim, J.J.; Kang, S.K.; Schatten, G.; Hwang, W.S. Dogs cloned from adult somatic cells. Nature, 2005, 436: 641-646.

[46] Mello, M. R. B. Clonagem em Bovinos. In: Palhan, H.B. (ed.) Reprodução em Bovinos. Fisiopatologia, terapêutica e Biotecnologia. 2nd Ed. LF livros, 2008. p.225-233

[47] Foote, H.R. Historical Perspective. In: Cibelli, J.; Lanza, P.R.; Campbell, K.H.S.; West, M.D. (ed.) Principles of Cloning. Academic Press, 2002. p. 3-14

[48] Bordignon, V.; Henkes, L.E. Clonagem Animal: a busca da amplificação de cópias geneticamente modificadas. In: Collares, T. (ed.) Animais Transgênicos. Princípios e métodos. São Carlos, Suprema, 2005. p137-166.

[49] Vajta, G.; Lewis, I.M.; Trounson, A.O.; Purup, S.; Maddox-Hyttel, P.; Schmidt, M.; Pedersen, H.G.; Greve, T.; Callesen, H. Handmade somatic cell cloning in cattle: analysis of factors contributing to high efficiency in vitro. Biology of Reproduction, 2003, 68: 571-578.

[50] Transgenic Animals http://en.wikipedia.org/wiki/Transgenic_animal#Transgenic_animals (accessed 3 September of 2012).

[51] Genetically modified mouse http://en.wikipedia.org/wiki/Genetically_modified_mouse. (accessed 1 September of 2012).

[52] Trangenesis history http://www.transgenicmouse.com/transgenesis-history.php. (Accessed 1 September of 2012).

[53] Nagy, A.; Gertsenstein, M.; Vintersten, K.; Behringer, R. Manipulating the mouse embryo. A laboratory manual, 3rd ed. Cold Spring harbor Laboratory Press, 2003.

[54] Houdebine, L.M. Métodos de gerar animais transgênicos e controle da expressão gênica In: Collares, T. (ed.) Animais Transgênicos. Princípios e métodos. São Carlos, Suprema, 2005. p 81-113.

[55] Montoliu, L.; Lavado, A. Animais transgênicos na biologia, na biomedicine e na biotecnologia. In: Collares, T. (ed.) Animais Transgênicos. Princípios e métodos. São Carlos, Suprema, 2005. p 114-136.

[56] Williams RS, Wagner PD. Transgenic animals in integrative biology: approaches and interpretations of outcome. J Appl Physiol. 2000 Mar;88(3):1119-26.

[57] Sutovsky, P. Somatic Cell Nuclear Transfer, Springer, 2007, 591.

[58] Williams, R. S.; Wagner, P. D. Transgenic animals in integrative biology: approaches and interpretations of outcome J. Appl. Physiol, 2000, 88: 1119–1126.

[59] Gordon I. Reproductive Technologies in Farm Animals: CABI; 2005.

[60] Houdebine LM. Cloning by numbers. Nat Biotechnol. 2003;21(12):1451-2.

[61] Yanagimachi R. Cloning: experience from the mouse and other animals. Mol Cell Endocrinol. 2002;187(1-2):241-8. 62.

[62] Miyashita N, Shiga K, Fujita T, Umeki H, Sato W, Suzuki T, et al. Normal telomere lengths of spermatozoa in somatic cell-cloned bulls. Theriogenology. 2003;59(7): 1557-65.

[63] FOX MW. Genetic Engineering Biotechnology 1: Animal Welfare and Environmental Concerns. Applied Animal Behaviour Science. 1988;20:83-94. Epub 94.

[64] Frewer LJ, Howard C, Shepherd R. Public concerns in the United Kingdom about general and specific applications of genetic engineering: risk, benefit, and ethics. Sci Technol Human Values. 1997;22(1):98-124. PubMed PMID: 11654686. eng.

[65] Hilgartner S. The Dominant View of Popularization: Conceptual Problems, Political Uses. Social Studies of Science. 1990;20(3):519-39. Epub 539.

[66] Barth WL. Engenharia Genética e bioética .Rev. Trim., 2005; v.35(149): 361-391.

Section 4

Ethics and the Role of Media in Reporting Controversial Issues in Biotechnology

Chapter 5

The Presentation of Dolly the Sheep and Human Cloning in the Mass Media

Miguel Alcíbar

Additional information is available at the end of the chapter

http://dx.doi.org/10.5772/56161

1. Introduction

He watched her drift away, drift with her pink face warm, smoothas an apple, unwrinkled and colorful. She chimed her laugh at everyjoke, she tossed salads neatly, never once pausing for breath. And thebony son and curved daughters were brilliantly witty, like theirfather, telling of the long years and their secret life, while theirfather nodded proudly to each.("The Long Years", Martian Chronicles, Ray Bradbury)On the next day, John saw Jesus coming toward him,and so he said: "Behold, the Lamb of God. Behold, he whotakes away the sin of the world".(John 1, 29)

1.1. Cloning as a media phenomenon

On 22 February 1997, the media covered the announcement of the birth of Dolly the sheep, the first mammal in history to be cloned from an adult cell. The animal's cloning by Ian Wilmut and his colleagues at the Roslin Institute, close to Edinburgh, rekindled a latent issue in popular culture: Is the cloning of human beings also possible? Dolly was the living proof that the images depicted in science-fiction literature and films could imminently become a disturbing prospect.

The media coverage of this story can be explained by the varied and complex implications of the human application of the *nuclear transfer*, cloning technique employed by Wilmut and his team to achieve the amazing feat of Dolly. Therefore, the media differentially framed the risks and benefits of human cloning. They trumpeted cloning as a means of curing a wide range of diseases and as a cheap and safe method of producing food en masse. But, above all, the media highlighted those applications of cloning that might potentially violate human nature.

According with [1], the author understands "human cloning" as the creation of a human embryo, whether for producing stem cells for biomedical purposes or for the gestation of a foetus and subsequent birth of a baby. Generally speaking, the media treat human cloning in

INTECH
open science | open minds

© 2013 Alcíbar; licensee InTech. This is a paper distributed under the terms of the Creative Commons Attribution License (http://creativecommons.org/licenses/by/3.0), which permits unrestricted use, distribution, and reproduction in any medium, provided the original work is properly cited.

a discriminatory way: if its techno-scientific applications and potential benefits for biomedicine are emphasized, it is known as "therapeutic cloning"; if, on the other hand, the discourse is keyed to human reproduction, with all the disturbing scenarios that this conjures up, it is called "reproductive cloning". However, this differential treatment of human cloning is clearly a rhetorical strategy for disencumbering certain manipulations of human embryos for research or therapeutic purposes of their negative connotations. This is due to the fact that the media generate a great deal of apprehension on framing human cloning as a regenerative or reproductive process. Nevertheless, the dichotomy is not at all clear. On the one hand, therapeutic cloning is based on a reproductive technology (nuclear transfer) and, on the other, reproductive cloning can be regarded as a therapeutic procedure for treating, for instance, infertility, according to the *in vitro* fertilization model.

Since the presentation of Dolly, the media have consolidated human cloning as a feasible "scientific fact". The chronological milestones that have contributed to this can be summarized as follows:

- **February 1997**. The media covered the official presentation of Dolly at the Roslin Institute.
- **January 1998**. The American physician Richard Seed made a controversial statement about his intention to clone a human being [2].
- **November 2001**. The *Journal of Regenerative Medicine* published a paper by scientists working for the biotechnical company Advanced Cell Technology (ACT), in which they claimed to have cloned a (six-cell) "human embryo" [3].
- **June 2002**. The Italian gynaecologist Severino Antinori announced that the first cloned baby had already completed 14 weeks of gestation [4].
- **December 2002**. Brigitte Boisselier, spokeswoman for the International Raëlian Movement and director of Clonaid, convened a press conference to announce the cloning of a girl called Eva [3, 4, 5].
- **January 2004**. Dr. Panos Zavos announced before the press in London that he had implanted a recently cloned human embryo in a sterile woman [6].
- **February 2004**. Dr. Woo Suk Hwang, a practically unknown South Korean researcher, claimed to be the first person to clone a human embryo and obtain stem cells from it, thus stepping into the international spotlight. In less than two years, from being an anonymous researcher Hwang became a national hero, which was to be his undoing since at the end of 2005 his research turned out to be a monumental hoax [7, 8].

The cloning of mammals and, above all, the possibility of applying cloning techniques to human beings, is therefore one of the most important public techno-scientific controversies of the turn of the century [5, 9]. Cloning is a media phenomenon because it provokes mixed reactions among different sectors of society. The controversial aspects of cloning range from those that are purely technical and its potential applications in the fields of biomedicine, livestock breeding, and crop farming, to ethical and faith-based moral issues, through those touching on the control mechanisms and legal regulation that these techno-scientific practices call for. Nevertheless, the press addressed these aspects in a biased way, magnifying some and minimizing others.

The focus was basically placed on the culmination of Dolly as a product and the social consequences of the experiment, rather than on its technical details. The information published about Dolly was more a catalyst of latent social fears than a driver for the pedagogical dissemination of scientific knowledge. This deep-rooted habit of journalism is usually heavily criticized by scientists, who dub it as "dumbing down" and hence a distortion of scientific research. However, techno-scientists aware of the power of the media as regards reach and publicity have exploited journalists for their own personal gain on quite a few occasions [10, 11]. The benefits that scientists and techno-scientific companies obtain from press coverage range from greater professional prestige and social legitimization to greater financial gain. It is with good reason that the extraordinary amount of publicity that Dolly obtained led to a climb of 65% in the share price of PPL Therapeutics, the company sponsoring the experiment, on the London Stock Exchange, just three days after the announcement had been made [12].

The cloning of Dolly is a significant example of what is called the "mediatisation of science" [13, 14, 10]. Unlike other techno-scientific breakthroughs, whose technical details are usually made known via formal channels, the cloning of mammals was disseminated via the mass media. This meant that the strictly scientific side of cloning had a lesser impact on public debate, while its social repercussions and disturbing future scenarios were indeed magnified [15, 16, 17].

This can be explained by taking a look at two interrelated factors: 1). Nowhere in the paper published in *Nature*, in which Wilmut and his team describe their experiments on the viability of offspring derived from foetal and adult mammalian cells, is there a mention to cloning and, even less, human cloning [18]; and 2). The possibility of applying the technique used by the Scots research team to human beings is based solely on a unique and fortunate animal experiment, not without its controversial aspects (see Note 3).

Despite these restrictions, the exploitation of Dolly as a media phenomenon – that is, the amplification of social controversy – was due in part to the fact that its rapid creation was related to given cultural suppositions, steering the debate towards the hypothetical although plausible field of human cloning. As a result, Dolly mobilized different social actors that, with divergent interests and arguments, shifted cloning into the realm of human testing and the ethical issues that these tests might raise. Such an extrapolation stirred up, undoubtedly, an overwhelming fear of automated replication, mass production and the loss of individuality, all of which belong to the recurring imagery of popular culture as regards cloning [19].

The fact that the journalists knew a little beforehand that the paper of Wilmut and his colleagues had been accepted for publication in *Nature*, one of the scientific community's most widely circulated journals, accorded the research a tacit legitimacy [15]. In any case, the lack of a critical attitude towards information coming from a scientific source, and the fact that the journalists in question probably did not even read the original paper in *Nature*[1], in addition to the lack of independent inquiry into the validity of the experiment, seem to be factors crucial to understanding why they saw fit to air their own opinions on the significance of the cloning of Dolly.

These disturbing journalistic interpretations triggered social and political pressure in favour of banning cloning [21, 22]. As will be seen, the statements of experts might also have contributed significantly to shifting the focus from animal cloning to the human kind, with the hope of safeguarding the former from restrictions. Neither was there any debate about (non-genetic) environmental influences on the development of the clone [23], nor – and this happens to be one of the most controversial points – on the differentiation state of the mammary cell used by Wilmut [24].[2] The cloning of Dolly was not only unusual because of the extraordinary amount of media coverage that it received, but also because the controversy – unlike other cases such as that of cold fusion [25, 26] – was not about scientific facts (although there were news stories that varied with regard to their representation of these facts), their interpretation, or even their implications for *per se* policies. The controversy was about how these facts affected ethical issues [27, 28].

Sticking to the reasons given by the Scots scientists and their sponsor PPL Therapeutics, the experiment that led to the birth of Dolly was conceived so as to develop lines of research on cellular differentiation and other basic aspects of cellular biology; in addition to opening up new ways of using cloning techniques in the field of biomedicine and livestock farming, with the commercial gain that its sponsors expected to obtain from such applications [15, 29]. However, the strictly scientific controversies, that is to say, those related to discrepancies in the interpretation of the data, the experimental protocols used, or the skill of the researchers, were practically ignored by the press. Furthermore, its application in the fields of medicine and livestock breeding were only taken into account in an advanced phase of the debate [30]. In contrast, as already mentioned, the debate focused on the ethical issues stemming from this new biotechnology and on the need to legislate on its application in human beings. An example of this can be found in the way that the press covered the success rate of nuclear transfer. Although the scant success rate of the method was the reason for dispute in scientific circles (Dolly was the sole successful result out of 277 previous attempts),[3] this was never a controversial point for the journalists covering the story. It was only mentioned so as to illustrate how immensely difficult it was to clone a mammal from a mature cell. This example shows that, generally speaking, the media ignore those technical details on which there is no expert consensus.

---

1 The first news about Dolly was published on 24 February 1997, despite the fact that the paper appearing in *Nature* was not published until the 27th. The story was brought to light by a scientific editor working for *The London Observer*, who obtained the information from a source other than *Nature*, thus technically breaking the embargo that the journal had placed on the information (see [15, 20]).

2 Authors such as [24] dispute whether or not the mammary cell used by Wilmut to clone Dolly was in fact an adult cell. Since Dolly was developed from a cell extracted from the mammary gland of a six-year-old sheep in its last three months of pregnancy, and that it is known that given the fact that the mammary glands of mammals increase in size during the final phases of gestation, it is permissible to deduce that some mammary cells, although technically adult, still behave in a highly labile way, or even in a similar fashion to embryonic stem cells. This situation would lead them to be regarded as undifferentiated cells and, therefore, totipotent. As Gould has indicated, maybe it is only possible to clone from unusual adult cells with a potential embryonic effect, and not from any cheek cell, hair follicle, or drop of blood that falls prey by mere chance to a mad photocopier.

3 The 277 attempts is the figure published in the press; however, in a paper originally published in *Science*, the success of the experiment conducted with these mammary cells was practically attributed to a miracle: 434 nuclear transfer tests failed, but not Dolly's [31].

2. The cloning of Dolly as a "scientific fact"

The media converted Dolly into a kind of totemic animal, a sign of the times (Figure 1). It became a popular symbol of the trangressive potential of *new genetics*, since it was thought that its creation had violated certain biological dogmas [32]. Although Dolly was the result of a "successful" one-off experiment, for the media the animal's birth represented the *irrefutable proof* that cloning by somatic cell nuclear transfer was not only feasible, but also that its application in human beings had ceased to be a futurist dystopia to become a dismal techno-scientific prospect.

On the basis of the Actor Network Theory (ANT), the sociologist from the University of Trento Federico Neresini has demonstrated the role played by the mass media in establishing the cloning of Dolly as a genuine "scientific fact" [5]. His conclusions are based on the analysis of 95 articles published in two of Italy's most widely read daily newspapers – *Il Corriere della Sera* and *La Repubblica* – during the apogee of the Dolly case, that is, from 22 February-10 March 1997.

Figure 1. Professor Ian Wilmut and Dolly (Source: Roslin Institute)

According to the ANT, "scientific facts" are such thanks to complex processes of translation within heterogeneous networks in which different actors negotiate, among other things, the ontological statute of those facts. If the network's main actors are capable of persuading the rest of the need for establishing certain pretentions of knowledge as "scientific facts," then it is possible to say that these can be socially implanted with success, at least temporarily. Although the ANT does not underestimate the fact that common sense has made us accustomed to distinguishing "scientific facts" from the context in which they are produced, it does not accept the dichotomy between science and society, which it regards as false, and looks upon this disjunction as being an effect of the social process rather than its starting point. For this reason, ANT sociologists talk about *hybrids*: Dolly can be considered as a good example of a hybrid, since it is impossible to exclusively classify it as a techno-scientific fact, social construction, or natural entity [33].[4] For Neresini, during the *chain of translations* Dolly, as a "scientific fact," shifted from one set of contexts to another so as to attract the attention of new and varied actors. This means that, in some way, the "scientific fact" can acquire different meanings for these new actors (hence, translation as betrayal and the hybrid notion as something impure and hazy), distinct from its meaning for the researchers responsible for the experiment. The latter's concern was basically to consolidate animal cloning, according to certain techno-economic criteria [5]. Neresini observes that during the first few days of debate in the Italian press, the network of actors spread, thus giving rise to the first translations. The objectives of these actors, other than being diverse, were also in some cases contradictory: to consolidate their own opinions about *in vitro* fertilization, to put the accent on its applications in the field of livestock breeding and experimental medicine, to limit scientific research, especially in the area of genetic engineering, or to avoid the risk of denaturalizing reproduction, with the consequent loss of human identity, among others. However, they all contributed to socially reinforcing the cloning of mammals from differentiated cells as a genuine "scientific fact". A clear example of translation was that made by the Catholic Church. The Church used the debate on the cloning of Dolly to strengthen its beliefs by reopening other collateral debates such as that of abortion, contraception or the social definition of "family"; way beyond the expectations of Wilmut and his team when they thought up and conducted the experiment. So, the ability of the main actors in a heterogeneous network consists in making diverse divergent aspirations converge in a common objective: in the case in hand, accepting the cloning of Dolly as an unquestionable "scientific fact".

It is interesting to note that the actors that opposed human cloning could not help but maintain the cloning of Dolly as a genuine "scientific fact", since they were not opposed to the "scientific proof" that Dolly represented, but precisely against the application in human beings of certain biological principles that had led to this achievement. The fact of cloning is taken for granted; what is rejected is human cloning, with arguments of an ethical (as in the case of Dolly) or techno-social-political nature (as in the case of the Raëlian movement in the Spanish daily *El País*; see [3]). Even the Catholic Church was interested in establishing the cloning of mammals from somatic cells as a "scientific fact", although neither with the aim of improving its own scientific reputation, nor that of defending freedom of inquiry (which, for obvious reasons, was indeed in the interests of the team led by Wilmut), but with the aim of condemning abortion and assisted reproduction techniques with scientific arguments, so as to reaffirm a certain family model (defined by Catholic morals as "natural") and to reclaim the authority of the Church as regards the definition of the meaning of "human being". So, public debate on the possible uses and/or consequences of the use of cloning techniques in human beings legitimized the issue as a "scientific fact", at least in the mass media world [5]. What is more, if for a limited core of experts the cloning of Dolly might have been technically controversial, the mass media actively contributed to constructing it publically as an indisputable fact, focusing on certain elements of the debate and excluding others. The media helped citizens, policy-makers, businessmen, and scientists accept the phenomenon of the cloning of Dolly as a genuine "scientific fact", each defending their own interests.

4 Franklin suggests regarding Dolly as a form of ownership. All forms of *ownership* are cultural inventions, and Dolly cannot only be regarded as a scientific invention, or as an ethical dilemma, but also as a cultural product.

It is widely known that media agenda setting had a powerful influence on political decision-making about cloning, both at an administrative and legislative level. In some countries, political reactions to the announcement of Dolly were quick and decisive, the majority of them coming before the publication of Wilmut et al.'s paper in *Nature*. This rapid political reaction suggests that establishing Dolly's cloning as a "scientific fact" and its possible applications in humans played a decisive role in the tone of the statements made and in the nature of the directives issued by the main official agencies (UNESCO, UN, EU, etc.) and world governments. Media coverage determined to a great extent the focus of policies on research that might affect the nature of human life. As it happens, for instance, the British government withdrew the funds assigned to Wilmut's research group [15].

Media agenda setting also had an influence on the political agenda. The mass media do not try to force people to think in a certain way, but they do indeed succeed in narrowing down the issues that in their opinion should concern the general public [34]. The hopes of the general public as to the potential future benefits of cloning, along with their fears about eventual malicious applications, imply that people accepted the cloning of Dolly as a well-established "scientific fact", giving legitimacy to the experiment conducted at the Roslin Institute.

Therefore, the role of the media in socially establishing the cloning of Dolly as an undisputable "scientific fact" was decisive, since they contributed to sustaining a heterogeneous network of actors that, by means of chains of translation, linked Dolly's cloning to other situations that the scientists responsible for the experiment had never even contemplated, explicitly at least, such as *in vitro* fertilization, the ontological statute of the human embryo, or the loss of individuality. Due to this, many other actors were prepared – for diverse reasons and with different objectives in mind – to be included in the debate and thus steer the discussion towards topics that already formed a part of the thematic agenda of the media. Dolly has at least two characteristics that make it ideal for arousing media interest. The first is that it has an identifiable name and image, and the second is that cloning has sufficient ingredients of attraction and repulsion so as to fit the type of stories told by the media. It awakens our collective imagination and affects our emotions on linking techno-scientific advances with images that are deeply rooted in popular culture. In this sense, it is important to take into account that the media are one of the main actors in the construction of heterogeneous networks in which identities, interests and facts are negotiated.

3. Nuclear transfer, techno-scientific biofantasies and the "exact copy myth"

As already mentioned, the announcement of the birth of Dolly was a major media event. During the whole of 1997 and part of 1998, the ethical debate centred on the possibility of applying the technique to human beings, grabbing the headlines in a number of newspapers and generating a significant amount of informative content and opinion [35]. The evolution of the social debate on cloning was clear in the Spanish context. Since the first days following the presentation of Dolly, representations based on science fiction and the fears stemming from these got the upper hand on the technical descriptions of the experiment. During 1997, the media presented the debate as an ethical and legislative problem, before bringing it in line, from 1998 onwards, with a discourse more akin to the biomedical applications of the novel method – tissue banks, organ transplants avoiding the problems of genetic rejection, or human reproduction (see Peralta quoted in [35]; [30]). In 1999, little was published about cloning, but from August 2000 onwards, with the British government's acceptance of the cloning of human embryos for therapeutic purposes, the ethical and legislative debate re-surfaced [36]. The declaration of the United States Congress of 1 August 2001, banning the use of human embryos for biomedical research purposes, as well as the statements made by the Italian gynaecologist Severino Antinori about his intention of cloning humans, rekindled the ethical debate on the boundaries of scientific research. Furthermore, in November of the same year, the company ACT announced that it had managed to clone a human embryo [37]. At the end of 2002 and the beginning of 2003, the announcement made by the Raëlians about the cloning of a healthy baby girl reopened yet again the debate on the boundaries of research and the need to legislate as regards these practices. In particular, in the Spanish daily *El País* cloning was presented as more of a scientific policy problem than an ethical issue. It called on policy-makers to clearly differentiate between reproductive cloning – ethically and technically reprehensible – and therapeutic cloning – necessary for combating certain degenerative diseases. It was hoped that the former would be banned and the latter promoted [3]. From this, it is clear that the ethical debate has always revolved around the need for enacting laws on the use of reprogenetic techniques.

The media frequently describe cloning as a procedure for obtaining "exact copies" from an original mould. As a result, cloning awakens public concern about genetic uniformity. However, the nuclear transfer technique generates, as it were, "more imperfect copies" than those represented by monozygotic twins, since these develop from the same fertilized ovum, while Dolly developed independently from the donor ewe [24, 38, 39]. Nevertheless, the press simplified the issue and used literary and film stereotypes present in popular imagery as a benchmark.

The technique used to clone Dolly is easy to understand. It involved introducing the nucleus of a somatic cell, taken from the udder of a white donor sheep, into an enucleated ovum (from whose nucleus all the genetic material had been previously removed) of a black-faced sheep, which behaved from this moment on as if it has been fertilized. With the fusion of the nucleus of the adult cell and the enucleated ovum by means of electrical discharges, a "reconstructed ovum" was obtained in laboratory conditions which was then implanted in a third sheep (also black-faced) which ultimately engendered Dolly (Figure 2).

Nuclear transfer is a *reprogenetic technology*, that is to say, a technology geared to the genetic reprogramming of the manipulated cell. In the strict sense of the word, Dolly is identical to the ewe that donated the mammary cell only in terms of nuclear genetic material, but clearly different with respect to the micro- and marco-environmental factors to which it was exposed (conditions depending on the uterus containing the embryo and the unique events making up the life history of each individual) (Peralta quoted in [35]).

On drawing upon social stereotypes, the media contribute to disseminating and publically establishing certain myths of a scientific origin in a continuous dialectic process of information

http://dx.doi.org/10.5772/56161

flow. The "exact copy myth" of cloning that threatens human uniqueness and individuality is, without doubt, a stereotype that the media use to simplify information and satisfy the rhetoric of emotions [40]. Appealing to the rhetoric of emotions is a very effective strategy if in addition it is reinforced by an efficient rhetoric of scientific rationality, which lends the discourse a sufficient level of credibility so as to defend politically-correct social attitudes.

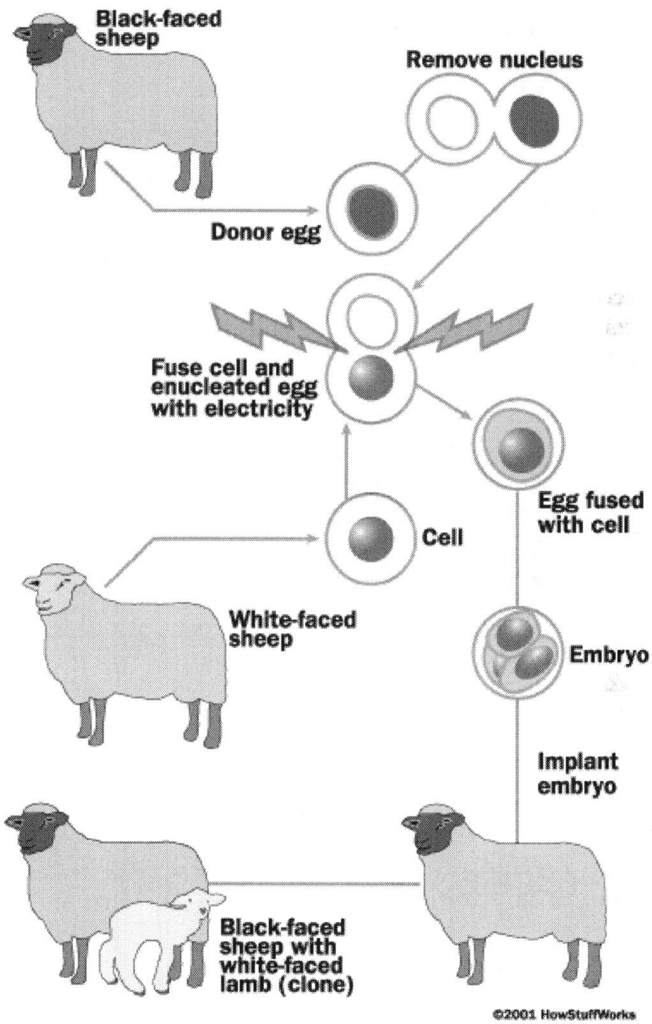

Figure 2. Nuclear transfer is the technique used to produce the embryo that resulted in the birth of Dolly the sheep (Source: HowStuffWorks).

Peralta (quoted in [35]) found that, just after the announcement of Dolly's birth, an initial effect of rejection of cloning was produced due to the way in which the media covered the story: the debate rapidly shifted to the ethical problems related to the possibility of cloning human beings (or parts of them) [5, 19, 30, 33]. According to Peralta, several factors contributed to conjuring up such a disturbing and, to a certain degree, perverse image of cloning. On the one hand, the continuous references to the diffuse symbolic imaginary created by science-fiction literature and films, above all, the futurist scene described by Aldous Huxley in *Brave New World* [41-44] and the technical madness of multi-cloning Hitler in both the book and film versions of Ira Levin's *The Boys from Brazil*. On the other hand, the photos and infographics included in news stories reinforced the "exact copy myth". Both factors worked synergistically.

A quick glance at some of the headlines of stories published in the Spanish press during the first week after the announcement of Dolly's birth might help to see how science fiction helped to evoke images, once seen as "terrifying fiction" and now, thanks to techno-scientific progress, as feasible:

- La oveja «Dolly» abre el camino para crear humanos en serie (Dolly the sheep opens the way to mass producing humans) (*El Periódico*, 24/02/97).
- La ciencia-ficción se convierte en realidad (Science fiction becomes reality) (*El Mundo*, 24/02/97).
- Dolly: entre animal y máquina (Dolly: half animal, half machine) (*El Mundo*, 25/02/97).
- La oveja «Dolly» resucita el fantasma de la clonación de seres humanos (Dolly the sheep resuscitates the spectre of human cloning) (*ABC*, 25/02/97).
- Dolly no fue la primera. *La literatura y el cine se adelantaron a la ciencia en la creación de clónicos* (Dolly was not the first. *Literature and films anticipated science in the creation of clones*) (*La Vanguardia*, 26/02/97).
- «Dolly» abre la puerta a la copia de personas muertas y congeladas (Dolly opens the way to copying dead and frozen people) (*El Periódico*, 28/02/97).
- Las ovejas clónicas convierten la ciencia-ficción en realidad (Cloned sheep makes a reality of science fiction) (*La Vanguardia*, 01/03/97).
- Las imposibles granjas para humanos (The impossible human farms) (*El Periódico*, 02/03/97).
- Frankenstein y su obra (Frankenstein and his work of art) (*El Mundo*, 02/03/97).

Infographics and photos also played an important role in giving the impression that Dolly was in all senses identical to the ewe from which the mammary cell - with which the sheep was cloned – was extracted (Peralta quoted in [35]). From then on, the visual representations of cloning publically established the false image of clones as being "exact copies" (Figure 3).

In the case of Australia's main newspapers, Alan Petersen arrived at similar conclusions: both the verbal information and the visual messages (including infographics explaining the process used to create Dolly) disseminated and reinforced the popular image of cloning as a kind of "Xerox" mechanism [15].

http://dx.doi.org/10.5772/56161

4. Cloning as a laboratory counterfeit and genetic determinism

One of the most alarming images that the mass media highlighted about cloning was the "loss of individuality". The idea that a cloned person is not a unique individual implies two very closely related assumptions [19]. The first is that however exact the copy (clone) is, it does not transcend its condition of "laboratory counterfeit". The spurious nature of the clone is identified with its illegal provenance. After announcing a five-year federal moratorium on human cloning, Bill Clinton, the then President of the United States, stated this perception very eloquently.[5] Indeed, a clone, as an illegitimate laboratory copy, is regarded as an unnatural entity, that is, artificial, and therefore its "production" is contrary to human dignity. The story in *Time* magazine, for instance, held that "Dolly does not merely take after her biological mother. She is a carbon copy, a laboratory counterfeit so exact that she is in essence her mother's identical twin" (10 March 1997, p. 62).

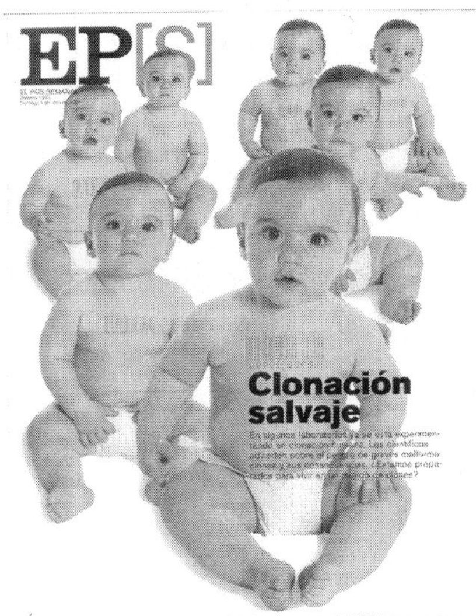

Figure 3. The photographic composition illustrating the report Clonación salvaje (Savage cloning) reinforces the "exact copy" myth, so frequent in popular representations of human cloning (Source: *El País Semanal* 1279, 1 April 2001).

The second assumption implies that the idea of loss of individuality is directly related to the first one. What is involved is the popular belief that genes determine *all* the characteristics of an individual. It is what is known as *genetic determinism*. Belief in genetic determinism leads one to conclude that the copy will be identical to its original, including its psychological

5 "What the legislation will do is to reaffirm our most cherished beliefs about the miracle of human life and the God-given individuality each person possesses. It will ensure that we do not fall prey to the temptation to replicate ourselves at the expense of those beliefs [...]. Banning human cloning reflects our humanity. It is the right thing to do. Creating a child through this new method calls into question our most fundamental beliefs" (Clinton quoted in [19]).

attributes, although its social status is of a lower rank. In [45, 46] has shown that the media depict genes and their iconic representation in a regular and ubiquitous way, emphasising their role in health, human behaviour and its diversity. In popular culture, genes have emerged as the panacea that provide simple, irresistible and apparently scientific answers to questions that are as complex as they are eternal: the cause of good and bad, the foundations of moral responsibility, and the nature of human relations. For these authors, media representations of genes express a genetic essentialism that favours biologically determinist and socially discriminatory public attitudes.

In connection with the representation of genes as omnipresent and ubiquitous entities, [47] points out that in the 1990s preference was given in the press to determinist representations that associated a certain gene to a disease or a human behaviour. On many occasions, the headline is determinist and the body of the news item is not, thus producing the so-called *framing effect*. According to this technique, the headline of the story substitutes the content (it frames it, so to speak), because few people read the whole story. Even though the body of the text contains non-determinist information, the headline is so powerful that its effect "frames" the interpretation of the reader, who tends to regard the information as a whole as determinist. In the case of Dolly, the press rarely mentioned the influence of non-genetic (environmental) factors or that of multi-factor genetic interactions as causes of the phenotypic features of the clone [23]. News stories with a determinist headline and body of the text were more commonplace, although those with a determinist headline and a body of the text containing non-determinist references were also published, although less frequently.[6] The latter is what the author has coined as "headline-body dissymmetry", a relatively common phenomenon in scientific journalism covering genetics. Its most evident effect is the dissemination of paradoxical information: while the headline has been written according to deterministic criteria, the story's content tries to depict genes as not being totally responsible for the characteristics of an individual, but rather the latter is a result of a complex multi-factor interaction where genetics and the environment act in a synergetic way.

5. The media framing of human cloning and its associated metaphors/ images

As has been seen, the debate on cloning and genetic engineering is strongly influenced by fictional narratives and literary and film stereotypes. These products of popular culture represent in turn a hotbed for creating multiple images and metaphors, which are then widely used in media debates. Dolly aired what popular culture had already successfully exploited in Hollywood films, television series and best-sellers. Therefore, cloning as a possibility, above

6 The following story published on the front page of the Spanish newspaper *El Mundo* (24/02/1997) is a good example of this: La ciencia logra «fotocopiar» por primera vez a un mamífero vivo (Science manages to "photocopy" a live mammal for the first time). The body of the text contains phrases such as the following: "It is nothing less than an exact *genetic photocopy* of another sheep"; "With a sole mammary cell from an adult sheep, these Scots researchers have managed to *produce* another identical sheep". There is not one reference to environmental factors, but rather the accent is put on the powerful influence of genes in *determining* that the cloned sheep is *identical* to its "original version".

all with a perverse end in mind, had attracted the attention of the general public long before Wilmut and his team presented Dolly and their achievement was submitted to public opinion. Although the media debate was first channelled towards ethical and legal issues, other interpretive frames were used afterwards.

The intrinsic relationship between the media and their audiences is a complex phenomenon of which media scholars do not have a thorough understanding as yet [48]. However, the mediation role seems to clearly indicate that the media reinterpret events, using certain structures, parameters and values, which ultimately cater to specific interests and certain conceptions of reality. Thus, the treatment of information is constrained both by internal factors (psychosocial features of communicators, professional routines, editorial viewpoints, etc.) and by those of an external nature (far-reaching ideological frameworks, cultural myths and stereotypes, economic interests of media corporations, audiences, etc.) [49].

The techno-scientific issues covered by the media are subject to these constraints, since they are coded on the basis of ideological criteria, news value and cultural norms [50, 51]. On considering that the media represent one of the chief information sources for citizens and that public support is frequently a necessary condition for implementing some or other policy, media content becomes a critical component of the interactions between citizens and politicians.

On the conceptual basis of the framing theory, it is possible to identify groups of metaphors that function within specific media frameworks. Framing is the act of emphasizing certain aspects of an event (and minimizing others) so as to allow the audience to interpret and contextualize the information by making it more understandable [48, 52]. In other words, framing is to define certain issues – generally by the elites – for public consumption, and to disseminate these definitions by means of the mass media [53]. The media are exceedingly relevant actors in framing techno-scientific controversies with social, political, economic and ethical implications. To start with, it may be helpful to adopt the definition of framing put forward by [54], which has been most lauded in the field of communication studies:

To frame is to select some aspects of a perceived reality and make them more salient in a communicating text in such a way as to promote a particular problem, definition, causal interpretation, moral evaluation, and/or treatment recommendation for the item described.

Therefore, **frames** *define problems* – by determining what a causal agent does and at what cost and benefit, generally measured in terms of common cultural values; they *diagnose causes* – identifying the forces giving rise to the problem; they *make moral judgements* – evaluating causal agents and their effect; and they *suggest remedies* – offering and justifying ways of addressing the problem and predicting their probable impact.

Based on the work of several authors that have studied the application of framing to techno-scientific issues [55-58], the following media frames to human cloning and several examples of their associated metaphors/images should be considered. It is important to note that some frames have a positive valence (i.e., promise, progress, economic prospects), and others a negative valence (i.e., ethical, Pandora's box):

1. *Promise*: usually referring to developments that will have significant consequences on how people live, eat, and view healthcare. "Rhetoric of future benefits".

"[la clonación]... tiene como objetivo lograr animales, que actúen como verdaderas fábricas vivas de drogas y proveedores de órganos susceptibles de ser trasplantados a seres humanos" ("The aim [of cloning] is to produce animals that act like authentic living manufacturers of medicines and suppliers of organs susceptible to being used for human transplants") (*ABC*, 07/03/1997).

2. *Progress*: celebrating new developments, breakthroughs; direction of history; conflict between progressive/conservative-reactionary.

"Lo que cabe esperar de los responsables políticos es que se actúe diligentemente contra los intentos irresponsables de fotocopiado de bebés y, a la vez, se proporcione un apoyo decidido a las técnicas de clonación que sí tienen un fuerte interés biomédico" ("What is expected of policy-makers is that they take action against the irresponsible attempts to Xerox babies and, at the same time, strongly support cloning techniques that do indeed have highly interesting biomedical applications") (*El País*, 07/01/2003).

3. *Economic prospects*: economic potential; prospects for investment and profits; R&D arguments;

"Según los científicos, los ganaderos podrían beneficiarse de esta técnica al conseguir animales clónicos a partir de otros animales adultos de sus ganaderías que hubieran demostrado ser más productivos y resistentes a las enfermedades" ("According to scientists, stockbreeders could benefit from this technique so to obtain cloned animals from other adult animals forming a part of their livestock which have proved to be more productive and resistant to disease") (*La Vanguardia*, 24/02/1997).

4. *Ethical*: calling for ethical principles; thresholds; boundaries; distinguishing between acceptable/unacceptable risks in discussions on known risks; dilemmas.

"La modificación genética para evitar enfermedades será aceptada mucho antes que la destinada a *mejorar* cualidades de los hijos como la forma física o la inteligencia" ("Genetic modification for preventing disease will be accepted long before its use for improving the characteristics of children such as physical fitness or intelligence") (*El País*, 12/01/2003).

5. *Pandora's box*: calling for restraint in the face of unknown risks; warnings prior to the opening of floodgates; unknown risks as anticipated threats; catastrophe warnings; disturbing future scenarios related to science fiction.

See all the examples included in the chapter that correspond to this frame.

6. *Nature/nurture*: environmental versus genetic determination; inheritance issues.

"Con una sola célula de las glándulas mamarias de una oveja adulta, estos investigadores escoceses han logrado *fabricar* otra oveja idéntica" ("With a sole mammary cell from an adult sheep, these Scots researchers have managed to *produce* another identical sheep") (*El Mundo*,

24/02/1997). There is not one reference to environmental factors, but rather the body of the text represents cloning as a model of genetic determinism.

7. *Public accountability*: calling for public control, participation, public involvement; regulatory mechanisms; private versus public interests.

Los pioneros de la clonación advierten que la técnica sería aplicable en humanos en dos años. *El doctor Wilmut pide normas internacionales para evitar esta posibilidad* (The pioneers of cloning warn that the technique could be applied to humans in two years. *Dr Wilmut calls for an international regulatory framework so as to avoid this possibility*) (*ABC*, 07/03/1997).

8. *Globalization*: calling for global perspective; national competitiveness within a global economy

"La nación que no quiera subirse al tren del progreso está condenada a ser un país de tercera división" ("Nations that miss the train of progress will condemn themselves to being third rate countries") (*ABC Cultural*, 07/03/1997).

9. *Freedom of inquiry*: science vs. applied science or technology; value free science; neutrality of science.

"Quien adultera la ciencia no es el científico sino los mercaderes oportunistas que transforman la plusvalía de la ciencia en una moneda de cambio podrida de intereses ajenos a la mentalidad científica. Por lo tanto, no es la ciencia quien precisa ser regulada, sino los traficantes del progreso" ("Those who adulterate science are not the scientists themselves but opportunist merchants that transform scientific benefits into a bargaining chip whose interests go against everything that science stands for") (*ABC Cultural*, 07/03/1997).

This wide range of frames suggests that the media debate on human cloning, which began with the presentation of Dolly, was, and still is, complex and multifactorial. Along these lines, it is interesting to highlight that certain metaphors can have different meanings, depending on the context in which they are used [59]. For instance, there is the metaphor that identifies cloning with a Xerox mechanism producing perfect copies. For those that interpret cloning within a frame with a positive valence, clones are useful products and, therefore, desirable. For those that interpret cloning within a frame with a negative valence, cloning is to produce a copy that is contrary to the essence of human beings, opening the way to manipulation and totalitarian control. Therefore, any attempt at cloning a human being or his/her parts would be regarded as a reprehensible act.

6. The representation of human cloning as an ethical problem: The boundaries of scientific research

All the studies conducted on the media coverage of Dolly seem to coincide in pointing out that the press represented cloning as an ethical problem in urgent need of legal regulation [5, 15, 19, 23, 27, 28, 33]. Indeed, the mass media gave priority to ethical problems stemming from the possibilities opened up by the cloning of a mammal and created virtual scenarios to fuel public concern about human cloning. It is logical to presume that such a glut of information affected policies and social attitudes towards cloning to a greater extent than academic debate on bioethics would have been able to achieve on its own.

The representation of cloning as a fundamentally ethical problem revolves around three interconnected issues: 1). *The loss of human uniqueness and individuality;* 2). *The (nearly always perverse) motivations for cloning;* and 3). *The fear of irresponsible scientists or science out of control* [19].

6.1. Loss of human uniqueness and individuality

One of the greatest concerns shown by the media was the alleged loss of uniqueness, as a consequence of the clone's spurious nature. A loss of uniqueness leads inevitably to that of human identity. Among other significant examples published in the North American press, Hopkins points out that the photo appearing on the front page of *Time* magazine (10 March 1997) showed two large identical adult sheep against a background of around 30 small copies, with the caption: **Will There Ever Be Another You?** (Figure 4). The inside cover page talked about cloning as a "Xerox" mechanism, and the photomontage used as an introduction to the main body of the text depicted a fruit machine dispensing identical people. A last photo showed several identical human bodies coming out of a test tube.

For Hopkins, these visual images transmit a provocative message that clones are exact denaturalized copies, while the body of the text strives to clarify that clones are not in fact exact copies, that is to say, to explain the inconsistency of arguments based on genetic determinism ("headline-body dissymmetry" and "graphics-body dissymmetry"). It is interesting to point out, as Hopkins himself suggests, that journalistic commentaries that try to explain and clarify erroneous essentialist interpretations do not have a clear pedagogical purpose as regards the genetic basis of human behaviour, but rather try to persuade readers that their fears about the loss of uniqueness are unfounded. Therefore, it has been observed that the media exaggerate and mitigate, simultaneously, concerns about the assumption that a clone, as an "unnatural" copy, prejudices human dignity. Hopkins asks himself whether the dominant message of the media about the loss of uniqueness is not a manifestation of the American people's peculiar emphasis on individualism, for which reason he suggests that comparative studies be conducted in other countries with different values and beliefs. The author speculates putting forward the hypothesis that in the press of other countries this obsession with individualism would not occur. In this sense, the only indicative study to date is that conducted by Neresini on the Italian press [5]. In Italy, there was also concern about the loss of individualism, although not to the obsessive extent that Hopkins sees in North American press coverage.

6.2. Motivations for cloning

In an attempt to assess the market that human cloning could generate, the mass media have imagined multiple possibilities and scenarios that would require cloning to reach certain goals. Speculation on hypothetical future uses of cloning cannot be censured, but it seems that their influence on the public image of cloning is by no means negligible, especially when such virtual scenarios are presented as perverse and are morally assessed. These hypothetical examples find their way into the collective conscience, acquiring a certain dose of credibility [19]. Before such scenarios actually occur, people already have a more or less detailed idea of the motivations that others might have to resort to cloning. In order of appearance in the media, Hopkins has detected the following:

Figure 4. Cover of *Time* magazine (10 March 1997)

The Megalomaniac. This motivation stems from the images projected by science-fiction literature and films. For instance, in *The Boys from Brazil* an attempt is made to multiply clone Hitler so as to perpetuate Nazi ideology. Woody Allen's futurist satire *Sleeper* revolves around the desire to clone an evil political leader using his nose. In *Jurassic Park*, terrified innocent people flee from the attacks of hungry *Tyrannosaurus Rex* clones, created so as to entertain visitors to a theme park. But these references to science fiction only illustrate "hypothetical scenarios". However, *Time* magazine toyed with the possibility of an eccentric millionaire that has never wanted to have children, but now, thanks to cloning, can have a child that not only bears his name but also his own genetic code. The magazine concluded: "Of all the reasons for using the new technology, pure ego raises the most hackles." (10 March 1997, p. 70). Despite having previously rejected genetic determinism, *US News & World Report* also clearly echoed the idea that a megalomaniac might decide to immortalize his or herself by cloning an "heir" (10 March 1997, p. 60).

The Replacement Child. This is the motivation of couples that want to "replace" a dying child. Along these lines, the benchmark for the global press industry *The New York Times* asked readers to consider "the case of a couple whose baby was dying and who wanted, literally, to replace the child" (24 February 1997: B8). On raising the issue of desperate situations such as this in such a naïve way, the media create paradoxes and myths. Implicit or explicitly, the "replacement" of a child implies that the cloned child will possess all the characteristics of the child being replaced, which contrasts with the simultaneous opinions of scientists and experts in ethics arguing against genetic determinism. It is important to note that the media transmit a negative image of these couples: they are people with psychological disorders, egoists and incapable of accepting death. A curious point is that they never align these motivations with those of other parents that, faced with the loss of a child, decide to have another, or even with the most common all for having children which is none other than to make the parents' lives more rewarding. The motivations of the former are pathological; those of the latter normal or even commendable.

The Organ-Donor Cloners. The mass media also raise the possibility of certain individuals resorting to cloning their offspring or themselves so as cure themselves from a disease or to create a genetically compatible organ and tissue bank. For instance, *Time* magazine (10 March 1997) began its special report with the hypothetical case of a couple whose only daughter has leukaemia: "the parents, who face the very likely prospect of losing the one daughter they have, could find themselves raising two of her—the second created expressly to help keep the first alive" (10 March 1997, p. 67). This motivation is usually treated with suspicion, as in the report published in *The New York Times* on 1 March 1997.

The Last-Chance-Infertile-Couple. This is presented as the least objectionable motivation for cloning. Also the least controversial, it could be justified depending on the medical status or the degree of misfortune of the infertile couple in question. Cloning would be the last resort for these couples, after trying orthodox fertility techniques that have failed. In this way, cloning is tacitly regarded as a psychological and morally inferior reproduction method than others.

6.3. Fear of irresponsible scientists or science out of control

The mass media have not only reported on cloning in negative terms, but they have also emphasized its potential benefits for medicine, agriculture and livestock breeding – although such benefits are always juxtaposed with their dangers [19]. On occasions, scientists are reproved for wanting to "play God"; this implies seeing science as an activity that can provide answers to many important questions, although its intrinsic amorality can be dangerous. For

http://dx.doi.org/10.5772/56161

instance, the headline of the article, **Little Lamb, Who Made Thee?**, appearing in *Newsweek* (10 March 1997), seems to point to the intrusion of scientists in the sacred domain of the divine.

But the most interesting discussion that the media construed on the boundaries of scientific research is based on the secular fear of its achievements and the perception that these are relentless. While it is reaffirmed that science is dangerous and that cloning is a technique against which people should react and, consequently, reject if allowed, it is recurrently admitted that science is relentless and that human cloning is inevitable, only being subject to the restrictions imposed by refining techniques and methods [60]. The same article appearing in *Newsweek* stated that the creation of Dolly teaches us a clear lesson.[7] From all this it can be inferred that the media and the general public perceive science as a robust enterprise as regards its achievements, amoral by definition, relentless in its progress, and inevitable in the application of its knowledge. From this perspective, the legal regulation of the application of scientific knowledge is like trying to gate-keep in a world without fences. The boundaries of scientific research are always of a technical kind which, one day, will be surpassed by the scientists themselves, but never boundaries stemming from ethical or other kinds of non-scientific imperatives.

In his analysis of the coverage of the Australian press, Petersen defends an identical stance. In the first stories to be published, journalists used phrases and metaphors evoking a kind of social engineering and authoritarian control. An ambivalent image is implicitly found in these articles: a belief in the all-embracing power of science, but also mistrust with regard to the motivations of scientists and fear of the results of their research [15]. A deep fear of "immoral science" is evident in many of the news stories about the cloning of Dolly. As [19] concludes: "The collective message here seems to be that a brave new world is detestable, but may be unavoidable".

7. The role of scientists in the defence of freedom of inquiry and the neutrality of science

After examining in some detail how the media represent cloning as an ethical problem in urgent need of legal regulation, it is interesting to analyze the role played by scientists themselves in focusing concern on the use of the technique in human beings, encouraged by their desire to defend their right to conduct research. What Dolly raised with redoubled starkness was the eternal issue about the "boundaries of science", specifically, about what should be done and how to regulate scientific research. In an attempt to preserve freedom of inquiry and its merits – i.e., a "neutral research model" – so as to impede policy-makers from implementing generic bans, the experts also contributed to sparking media concern about human cloning [5].[8]

7 "Science, for better or worse, almost always wins; ethical qualms may throw some roadblocks in its path, or affect how widespread a technique becomes, but rarely is moral queasiness a match for the onslaught of science" (p. 59).

Scientists became obsessed with clearly differentiating between animal and human cloning. They believed that this strategy would allow them to divert the focus from animal cloning to the human kind, in such a way that the former would not be seen as the gateway to the latter. The idea was to channel criticism from politicians, church authorities, and expert in bioethics towards human cloning, thus freeing animal cloning from moral and legislative burdens. This shift of focus was accompanied by an efficient "rhetoric of future benefits": the development of research in the field of animal cloning is important because it represents a source of potential benefits for medicine and livestock breeding. The "future benefits" strategy pretends to avoid public and political rejection of cloning, thus contributing to its social acceptance and, therefore, its development, on maintaining its sources of funding.

So, the researchers involved in the creation of Dolly did not limit themselves to technical comments about the experiment, but were more interested in safeguarding freedom of inquiry and the funding that makes it possible from the intrusions of politicians, church authorities, experts in bioethics, and public opinion. It is interesting to note that in the British press the differences of opinion on human and animal cloning constituted one of the most solid lines of argument in the debate on the cloning of Dolly. For example, an enormous effort was made to separate the idea that humans *should* be cloned versus whether this was actually possible; that is to say, an ostensible effort was made to separate the correct from the feasible. Therefore, animal cloning should have been seen as a positive concept, regardless of the technical, ethical and moral issues that the more than reprehensible cloning of humans would raise.

For his part, [5], in addition to the strategy for clearly differentiating between animal and human cloning, points to two others that scientists used to defend themselves from the offensive unleashed by certain political and/or religious groups: 1). *Emphasizing environmental rhetoric*: the separation between animal and human cloning is underlined, emphasizing the importance of environmental factors, at the expense of genetic factors, in the shaping of human identity. The experts tried to transmit the idea that even in the unlikely event that a person was cloned his or her identity would be safeguarded, since it depends on the unrepeatable history of an individual's interactions with the environment. In short, they attacked genetic determinism. So as to bring to the fore that cloning would not mean a loss of individuality, scientists referred to monozygotic twins as genetically identical but different as regards their behavioural and personality traits, and 2). *Distinguishing basic science from technology*: in their pursuit to safeguard a neutral research model, some scientists drew a clear boundary between basic and applied science. They tried to establish clear boundaries between basic scientific knowledge and its applications (science/technology dichotomy), as well as between scientific

8 With respect to this, several days after the announcement of Dolly's birth, Wilmut himself referred to human cloning only to condemn it. This can be seen in the following headlines and bylines appearing in the Spanish press: **La ciencia ficción se convierte en realidad.** *La técnica utilizada en Escocia puede utilizarse con las personas, pero los autores dicen que sería antiético* (**Science fiction becomes reality**. *The technique used in Scotland can be applied to humans, but the authors state that this would not be ethical*) (*El Mundo, 24/02/97*); **Ventajas e inconvenientes de una oveja clónica. «No vemos razones clínicas para clonar seres humanos», ha dicho el artífice de Dolly** (**The pros and cons of cloning sheep**. *"We don't see any clinical reasons for cloning human beings,"* states the person responsible for Dolly) (*El País, 26/02/97*); **Los pioneros de la clonación advierten que la técnica sería aplicable en humanos en dos años.** *El doctor Wilmut pide normas internacionales para evitar esta posibilidad* (**The pioneers of cloning warn that the technique could be applied to humans in two years.** *Dr Wilmut calls for an international regulatory framework so as to avoid this possibility*) (*ABC, 07/03/97*).

knowledge itself and non-epistemic values (science/values dichotomy). By means of this strategy, scientists aspired to shake off the responsibility for the "bad" ends to which others might eventually put their basic research. Therefore, they not only wanted to fend off personal accusations, but also to configure science as an intrinsically neutral activity. Consequently, science could continue on its path without the need for ethical or legal limits.

Science as value free is a very weak line of reasoning. On the one hand, because those same scientists are the ones demanding an acknowledgement for themselves and for scientific research, which, without doubt, stems from its potential applications: treating diseases, developing new medicines, improving transplant techniques, increasing livestock production, etc. And on the other, because, in a strict sense of the word, these scientists are techno-scientists; in other words, experts tied to the market demands of the biotechnological companies at which they work, or which sponsor them, as was the case of the team that cloned Dolly. Research at these companies is geared to obtaining economically profitable results.

On attempting to use the benefit rhetoric in the mid- and long-term to justify their research projects, scientists perversely showed that in reality their activity is closely linked to its applications, be they positive or negative [5].

8. Conclusions

The public presentation of Dolly the sheep unleashed certain latent biofantasies in popular culture, since they had already been successfully exploited by literature and films, like for instance the loss of human individuality, the mass production of slaves, or eugenics.

Even though Dolly was, in the strict sense of the word, only identical to the sheep donating the mammary cell as regards nuclear genetic material, the media described the animal as an "exact copy". This science-fiction approach conjured up disturbing future scenarios, and contributed decisively to framing the discourse on cloning more as an ethical problem in urgent need of legal regulation than a techno-scientific issue.

The scientists involved in the cloning of Dolly invested quite a bit of time in trying to clearly distinguish animal cloning (correct, feasible, ethically irreproachable, and with both commercial and biomedical benefits) from the human kind (reprehensible, immoral and unacceptable because of its technical risks).

Since the media play a relevant role in constructing social reality and modelling the images that the general public has as regards science and technology, it is important to consider the frames that they use to achieve this. By means of these frames, they emphasize or minimize certain aspects of an event so as to allow for its interpretation and contextualization, thus making it easier for the audience to understand the information. In this framing process, the media use multiple resources, including myths, cultural stereotypes, images, and metaphors, so as to make the information more accessible to the audience. As has been seen in the case of human cloning, these resources have been used profusely, which is a good indicator of the importance of the social debate fuelled by Dolly's presentation.

However, the presentation of Dolly involved a varied network of social actors (scientists, biotechnological companies, experts in bioethics, religious authorities, policy-makers, citizens, etc.) that, each in their own way and on the basis of their specific interests, contributed to establishing the cloning of Dolly as a genuine "scientific fact" and, consequently, the extrapolation of reprogenetic techniques to humans as feasible.

Author details

Miguel Alcíbar*

Address all correspondence to: jalcibar@us.es

Department of Journalism I, University of Seville, Spain

References

[1] Haran, J, Kitzinger, J, Mcneil, M, & Riordan, O. K. Human Cloning in the Media. From science fiction to science practice. London and New York: Routledge; (2008).

[2] Gerlach, N, & Hamilton, S. N. From Mad Scientists to Bad Scientists: Richard Seed as a Biogovernmental Event. Communication Theory (2005). , 15(1), 78-99.

[3] Alcíbar, M. Human Cloning and the Raelians: Media Coverage and the Rhetoric of Science. Science Communication (2008). , 30(2), 236-265.

[4] Ingram-waters, M. C. Public fiction as knowledge production: the case of the Raëlians' cloning claims. Public Understanding of Science (2009). , 18(3), 292-308.

[5] Neresini, F. And man descended from the sheep: the public debate on cloning in the Italian press. Public Understanding of Science (2000). , 9(4), 359-382.

[6] Haran, J. Managing the boundaries between maverick cloners and mainstream scientists: the life cycle of a news event in a contested field. New Genetics and Society (2007). , 26(2), 203-219.

[7] Kruvand, M, & Hwang, S. From Revered to Reviled: A Cross-Cultural Narrative Analysis of the South Korean Cloning Scandal. Science Communication (2007). , 29(2), 177-197.

[8] Park, J, Jeon, H, & Logan, R. A. The Korean press and Hwang's fraud. Public Understanding of Science (2009). , 18(6), 653-669.

[9] Holliman, R. Media coverage of cloning: a study of media content, production and reception. Public Understanding of Science (2004). , 13(2), 107-130.

[10] Weingart, P. Science and the media. Research Policy (1998). , 27(8), 869-879.

[11] Nelkin, D. Selling Science. How the press covers science and technology. New York: W. H. Freeman and Company; (1995).

[12] Reibstein, L, & Beals, G. A Cloned Chop, Anyone? Newsweek Magazine, 9 March, (1997).

[13] Rödder, S, & Schäfer, M. S. Repercussion and resistance: An empirical study on the interrelation between science and mass media. Communications (2010). , 35(3), 249-267.

[14] Schäfer, M. S. From Public Understanding to Public Engagement. An Empirical Assessment of Changes in Science Coverage. Science Communication (2009). , 30(4), 475-505.

[15] Petersen, A. Replicating Our Bodies, Losing Our Selves: News Media Portrayals of Human Cloning in the Wake of Dolly. Body & Society (2002). , 8(4), 71-90.

[16] Kasperson, R, Jhaveri, N, & Kasperson, J. X. Stigma and the Social Amplification of Risk: Toward a Framework of Analysis. In Flynn J. Slovic P. Kunreuther H. (eds.) Risk, Media and Stigma: Understanding Public Challenges to Modern Science and Technology. London: Earthscan; (2001). , 9-30.

[17] Kasperson, R, Renn, O, Slovic, P, Brown, H, Emel, J, Goble, R, Kasperson, J, & Ratick, S. The social amplification of risk: a conceptual framework. Risk analysis (1988). , 8(2), 177-187.

[18] Wilmut, I, Schnieke, A. E, Mcwhir, J, Kind, A. J, & Campbell, K. H. Viable Offspring Derived from Foetal and Adult Mammalian Cells. Nature (1997). , 385-810.

[19] Hopkins, P. D. Bad copies: How popular media represent cloning as an ethical problem. Hastings Center Report (1998). , 28(2), 6-13.

[20] Leach, J. Cloning, controversy and communication. In Scanlon E., Hill R., Junker K. (eds.) Communicating Science. Professional Contexts. London and New York: Routledge & The Open University; (1999). , 218-230.

[21] Mayor Zaragoza FLa clonación humana. Introducción. Quark, Ciencia, Medicina, Comunicación y Cultura (1999).

[22] Kutukdjian, G. La clonación humana con fines reproductivos: Cuestiones éticas. Quark, Ciencia, Medicina, Comunicación y Cultura, (1999). , 15-51.

[23] Petersen, A. Biofantasies: Genetics and Medicine in the Print News Media. Social Science & Medicine (2001). , 52(8), 1255-1268.

[24] Gould, S. J. Dolly's Fashion and Louis's Passion. In Nussbaum M.C. Sunstein C. R. (eds.) Clones and Clones: Facts and Fantasies about Human Cloning. New York and London: W.W. Norton and Co; (1998).

[25] Lewenstein, B. Cold fusion and hot history. In Scanlon E. Hill R. Junker K. (eds.) Communicating Science. Professional Contexts. London and New York: Routledge & The Open University; (1999)., 185-217.

[26] Lewenstein, B. From fax to facts: Communication in the cold fusion saga. Social Studies of Science (1995)., 25(3), 403-436.

[27] Priest, S. H. (2001). Cloning: a study in news production", Public Understanding of Science, 10, , 59-69.

[28] Priest, S. H. (2001). The Cloning Story", en A Grain of Truth. The Media, The Public, And Biotechnology, Rowman & Littlefield Publishers, , 97-108.

[29] Fransman, M. Designing Dolly: interactions between economics, technology and science and the evolution of hybrid institutions. Research Policy (2001)., 30(2), 263-273.

[30] De Semir, D, & Adrover, T. La clonación vista por la prensa española. Utilización de la base de datos Quiral para análisis de casos. Actas del I Congreso sobre Comunicación Social de la Ciencia. Comunicar la Ciencia en el siglo XXI (Libro II), March 1999, Granada, Spain. Granada: Proyecto Sur de Ediciones; (2000)., 25-27.

[31] Pennisi, E, & Vogel, G. Transformar el ensayo de Dolly. Mundo Científico (2000)., 217-21.

[32] Franklin, S. Culturing Biology: Cell Lines for the Second Millennium. Department of Sociology, Lancaster University; n. d. http://www.comp.lancs.ac.uk/sociology/soc022sf.htmlaccessed 9 October (2008).

[33] Franklin, S. Animal Models: an anthropologist considers Dolly. Department of Sociology, Lancaster University; 1998. http://www.comp.lancs.ac.uk/sociology/soc022sf.htmlaccessed 9 October (2008). Paper presented in the Second Symposium of the European Network for Biomedical Ethics, Maastricht, The Netherlands.

[34] Rodrigo Alsina MLa construcción de la noticia. Barcelona: Paidós; (1989).

[35] De Semir, V, Revuelta, G, Roura, M, De Semir, D, & Androver, T. Informe Quiral 1998. Medicina y Salud en la Prensa Diaria. Barcelona: Fundació Privada Vila Casas, Observatori de la Comunicació Científica y Universidad Pompeu Fabra; (1999).

[36] De Semir, V, & Revuelta, G. Informe Quiral 2000. Medicina y Salud en la Prensa Diaria. Barcelona: Fundació Privada Vila Casas, Observatori de la Comunicació Científica y Universidad Pompeu Fabra; (2001).

[37] De Semir, V, & Revuelta, G. Informe Quiral 2001. Medicina y Salud en la Prensa Diaria. Barcelona: Fundació Privada Vila Casas, Observatori de la Comunicació Científica y Universidad Pompeu Fabra; (2002).

[38] Bruce, D. M. Polly, Dolly, Megan, and Morag: A view from Edinburgh on cloning and genetic engineering. Phil & Tech (1997)., 3(2), 37-52.

[39] Johnson, G. Soul Searching. In Nussbaum M.C. Sunstein C.R. (eds.) Clones and Clones: Facts and Fantasies about Human Cloning. New York and London: W.W. Norton and Co; (1998).

[40] Van Dijk, T. A. News as Discourse. Hillsdale, NJ: Lawrence Erlbaum Associates; (1988).

[41] Nerlich, B, Clarke, D. D, & Dingwall, R. Fictions, fantasies, and fears: The literary foundations of the cloning debate. Journal of Literary Semantics (2001). , 30(1), 37-52.

[42] Huxford, J. Framing the Future: science fiction frames and the press coverage of cloning. Journal of Media & Cultural Studies (2000). , 14(2), 187-199.

[43] Van Dijck, J. Cloning humans, cloning literature: Genetics and the imagination deficit. New Genetics and Society (1999). , 18(1), 9-22.

[44] Turney, J. Frankenstein's Footsteps. Science, Genetics and Popular Culture. New Haven and London: Yale University Press; (1998).

[45] Nelkin, D, & Lindee, S. The DNA Mystique. The gene as a cultural icon. New York: W. H. Freeman and Company; (1995).

[46] Nelkin, D, & Lindee, S. Del gen como icono cultural. Mundo Científico (1998). , 194-71.

[47] González Silva MDel factor sociológico al factor genético. Genes y enfermedad en las páginas de El País (1976-2002). DYNAMIS. Acta Hisp. Med. Sci. Hist. Illus. (2005). , 25-487.

[48] Mcquail, D. McQuail's Mass Communication Theory. Thousand Oaks, CA: SAGE; (2005).

[49] Shoemaker, P. J, & Reese, S. D. Mediating the message: Theories of influence on mass media content. White Plains, NY: Longman Publishers; (1996).

[50] Boykoff, M. T, & Boykoff, J. M. Balance as bias: global warming and the US prestige press. Global Environmental Change (2004). , 14(2), 125-136.

[51] Carvalho, A. Ideological cultures and media discourses on scientific knowledge: re-reading news on climate change. Public Understanding of Science (2007). , 16(2), 223-243.

[52] Entman, R. M. Projections of power: Framing news, public opinion, and U.S. foreign policy. Chicago, IL: The University of Chicago Press; (2004).

[53] Berinsky, A. J, & Kinder, D. R. Making sense of issues through media frames: Understanding the Kosovo crisis. Journal of Politics (2006). , 68(3), 640-656.

[54] Entman, R. M. Framing: Toward clarification of a fractured paradigm. Journal of Communication (1993). , 43(4), 51-58.

[55] Reinhart, H. C. Framing the Biotechnology Debate: A Textual Analysis of Editorials and Letters to the Editor in the *St. Louis Post-Dispatch*. Global Media Journal (2007).

[56] Priest, S. H, & Eyck, T. News coverage of biotechnology debates. Society (2003). , 40(6), 29-34.

[57] Nisbet, M. C, & Lewenstein, B. V. Biotechnology and the American media: The policy process and the elite press, 1970 to 1999. Science Communication (2002). , 23(4), 359-391.

[58] Liakopoulus, M. Pandora's Box or panacea? Using metaphors to create the public representations of biotechnology. Public Understanding of Science (2002). , 11(1), 5-32.

[59] Hellsten, I. Dolly: Scientific Breakthrough or Frankenstein's Monster? Journalistic and Scientific Metaphors of Cloning. Metaphor and Symbol (2000). , 15(4), 213-221.

[60] Clarke, B, & Salmon, B. Public Perspectives on Human Cloning. A Social Research Study. Medicine in Society Programme, The Wellcome Trust; 1998. http://www.wellcome.ac.ukaccessed 15 March (2011).